Warr, Suhr

Chemical Information
Management

© VCH Verlagsgesellschaft mbH, D-6940 Weinheim (Federal Republic of Germany), 1992

Distribution:
VCH, P.O. Box 10 11 61, D-6940 Weinheim (Federal Republic of Germany)
Switzerland: VCH, P.O. Box, CH-4020 Basel (Switzerland)
United Kingdom and Ireland: VCH (UK) Ltd., 8 Wellington Court, Cambridge CB1 1HZ (England)
USA and Canada: VCH, Suite 909, 220 East 23rd Street, New York, NY 10010-4606 (USA)

ISBN 3-527-28366-8 (VCH, Weinheim)   ISBN 1-56081-180-3 (VCH, New York)

Wendy A. Warr, Claus Suhr

Chemical Information Management

VCH

Weinheim · New York · Basel · Cambridge

CHEMISTRY

Dr. Wendy A. Warr
ICI Pharmaceuticals
Mereside
Alderly Park Macclesfield
Cheshire SK10 4TG
Great Britain

Dr. Claus Suhr
Berlinerstr. 3
6701 Dannstadt
Federal Republic of Germany

Published jointly by
VCH Verlagsgesellschaft mbH, Weinheim (Federal Republic of Germany)
VCH Publishers, Inc., New York, NY (USA)

Editorial Directors: Dr. Hans F. Ebel and Dr. Wolfgang Gerhartz
Production Manager: Myriam Nothacker

A CIP catalogue record for this book is available from the British Library

LOC Card No. applied for.

Deutsche Bibliothek Cataloguing-in-Publication Data:

Warr, Wendy:
Chemical information management / Wendy Warr ; Claus Suhr.
– Weinheim ; New York ; Basel ; Cambridge ; VCH, 1992
  ISBN 3-527-28366-8 (Weinheim ...)
  ISBN 1-56081-180-3 (New York)
NE: Suhr, Claus:

Preface

This book arose from an article written in 1989 and the first part of 1990 for Ullmann's Encyclopedia of Industrial Chemistry, Volume B1, pages 12-1 to 12-119. The chapter on patent information in that article has now been expanded and moved to form Part Two of this book. Changes to the other chapters, which now form Part One, have been minimal because we felt that currency and speed of publication were particularly important. We have managed to insert some very recent references by removing older ones, but some more general ones have to be mentioned here.

In the writing of Chapters 6 and 7 on Information Technology, notes from the Information Technology Primers run by the Association of Information Officers in the Pharmaceutical Industry proved very useful. Much of the material from these primers has recently been incorporated into a book [0.1]. A recent review on storage technology [0.2] should be read in conjunction with Chapter 6.

In Chapter 8 we have cited a well known directory of text retrieval software [8.80] but failed to mention a directory with a wider scope which is now into its fourth edition [0.3]. A lengthy review of online searching aids (front ends, gateways and other interfaces) has also appeared [0.4].

A recent tutorial article on spectral databases [0.5] is pertinent to Chapter 9.

There have been several new publications in the field of chemical structure handling [0.6]–[0.10], all of which could contribute to Chapter 10. Two of these [0.7], [0.8] deal with distributed computing environments and electronic publishing and therefore have relevance to other chapters. The book by Ash et al. will supersede an earlier one [9.1], [10.3] and the software directory [0.10] is more up-to-date, and has wider scope than the book cited at [10.149].

It is almost impossible for two authors alone to cover the wide range of topics in this book, so thanks are due to many people who helped in the writing and researching. Dr. Peter Willett provided a copy of the reading list used by students at the Department of Information Studies at Sheffield University and a bibliography and MSc Thesis on Chemical Structure Processing by Karen Libscomb. Lipscomb's short descriptions of Chemical Abstracts tautomerism and of HTSS are reproduced almost verbatim in Chapter 10, with apologies, but with compliments for their accuracy and succintness. Dr Willett and Dr John Barnard offered helpful advice on chemical structure systems. Dr David Bawden traced various useful references. He, Karen Blakeman, Dr Trevor Devon and Dr Willett are to be thanked for their contributions to the IT primers mentioned above. The Chapter on Business and Economic Information was essentially written by Gillian Swash and Elizabeth Walton of ICI Pharmaceuticals. Janet Ash wrote Chapter 9 using material gathered, or earlier published, by Dr Wendy Warr. A lecture on hypertext by Jane Martin of ICI

Paints was used in the compilation of Section 11.7. John Barber. Section 6.4 on Microcomputers was contributed by Dr. William Town. Derek Bowler, Graham Cousins, Madeline Gray and Pat Holohan of ICI Pharmaceuticals checked many of the references. We thank the Beilstein Institute, Chemical Abstracts Service, and the Institute of Scientific Information for checking relevant sections of this book.

The preparation of Part Two has been greatly supported by some colleagues and organizations. We should like to express our sincere thanks to BASF Aktiengesellschaft, Ludwigshafen (Rhein), and in particular to the staff of its Documentation Department who were extremely helpful in contributing relevant literature and useful advice, further to Philip Higham and Raymond Andary of the World Intellectual Property Organization, Geneva, for providing valuable printed material, and to the management of Derwent Publications Limited, London, for permission to reprint Figure 16.

Abbreviations Used in this Book

ACS	American Chemical Society
ADMIS	Agfa Document Management Information System
ADONIS	Article Delivery Over Network Information Systems
AGLINET	Worldwide Network of Agricultural Libraries
AGRIS	Agricultural Sciences and Technology
AI	Artificial Intelligence
AIMB	Analogy and Intelligence in Model Building
ANSI	American National Standards Institute
API	American Petroleum Institute
AQUIRE	Aquatic Information Retrieval
ARGOS	Automatically Represents Graphics of Chemical Structures
ARIPO	African Regional Intellectual Property Organization
ARS	Agricultural Research Service
ASCA	Automatic Subject Citation Alert
ASCII	American Standard Code for Information Interchange
ASIDIC	Association of Information and Dissemination Centers
ASTM	American Society for Testing and Materials
BASIC	Basel Information Center
BIOSIS	BioSciences Information Service
BLDSC	British Library Document Supply Centre
CAD/CAM	Computer-aided design and manufacture
CAMEO	Computer Assisted Mechanistic Evaluation of Organic Reactions
CAN/SND	Canadian Scientific Numeric Database Service
CAR	Computer-assisted retrieval
CAS	Chemical Abstracts Service
CASDDS	Chemical Abstracts Service, Document Delivery Services
CASE	Computer Assisted Structure Elucidation
CASE	Computer-aided software engineering
CASP	Computer-assisted synthesis planning
CASR	Chemical Activity Status Report
CASSI	Chemical Abstracts Service Source Index
CCC	Copyright Clearance Center
CCD	Charge coupled device
CCDC	Cambridge Crystallographic Data Center
CCITT	Comité Consultatif International Télégraphique et Téléphonique
CCRIS	Chemical Carcinogenesis Research Information System
CDA/DDIF	Compound Document Architecture/Digital Document Interchange Format
CDC	Centers for Disease Control
CD-I	Compact disc interactive
CDIF	Crystal Data Identification File
CD-ROM	Compact disc — read only memory
CESARS	Chemical Evaluation Search and Retrieval System
CHEMICS	Combined Handling of Elucidation Methods for Interpretable Chemical Structures
CHIRON	Chiral Synthon
CHRIS	Chemical Hazard Response Information System
CIN	Chemical Industry Notes
CIS	Chemical Information System
CISTI	Canadian Institute for Scientific and Technical Information
CJO	Chemical Journals Online
CLF	Current Literature File
CNMR	Carbon-13 NMR Spectral Search System
CNRS	Centre National de la Recherche Scientifique
CODATA	Committee on Data for Science and Technology
COM	Computer-output microfilm
COUSIN	Compound Information System
CPI	Chemical Patents Index
CPI	Conference Papers Index

CPSS	Chemist&'s Personal Software Series
CPU	Central processing unit
CRDS	Chemical Reactions Document Service
CROSSBOW	Computerized Retrieval of Organic Structures Based on Wiswesser
CRT	Cathode ray tube
CSD	Cambridge Structural Database
CSO	Central Statistics Office
CTCP	Clinical Toxicology of Commercial Products
DAP	Distributed Array Processor
DARC	Description, Acquisition, Retrieval and Correlation
DCA/DIF	Document Content Architecture
DDE	Dynamic Data Interchange
DECHEMA	Deutsche Gesellschaft für chemisches Apparatewesen
DIF	Drug Information Fulltext
DIMDI	Deutsches Institut für Medizinische Dokumentation und Information
DIP	Document Image Processing
DIPPR	Design Institute for Physical Property Data
DOC	Dictionary of Organic Compounds
DOE	Department of Energy
DRI	Data Resources Inc.
DVI	Digital video interactive
EAGLE	European Association for Gray Literature Exploitation
EBCDIC	Extended binary coded decimal information code
ECDIN	Environmental Chemicals Data and Information Network
ECTR	Extended Connection Table Representation
EI	Engineering Index
EIIA	European Information Industry Association
EINECS	European Inventory of Existing Chemical Substances
EIU	Economist Intelligence Unit
ENVIROFATE	Environmental Fate
EPA	Environmental Protection Agency
EPI	Electrical Patents Index
EPIC	Thermosalt Estimate of Properties for Industrial Chemistry
EPO	European Patent Office
EPOQUE	EPO Query system
ERIC	Educational Resources Information Center
EROS	Elaboration of Reactions for Organic Synthesis
ESA	European System of Integrated Economic Accounts
ESA/IRS	European Space Agency Information Retrieval Service
ESPRIT	European Strategic Program for Research and Development in Information Technologies
EUSIDIC	European Association of Information Services
F*A*C*T	Facility for Analysis of Chemical Thermodynamics
FAO	Food and Agriculture Organization
FDA	Food and Drug Administration
FIZ	Fachinformationszentrum
FREL	Fragment Reduced to an Environment which is Limited
FRSS	Federal Register Search System
FT-IR	Fourier transform infrared
GAIIA	Global Alliance of Information Industry Associations
GCL	GENIE Query Language
GIABS	Gastrointestinal Absorption
GREMAS	Generic Retrieval by Magnetic Tape Search
GRUR	Gewerblicher Rechtsschutz und Urheberrecht
GUI	Graphical User Interface
HCI	Human — computer interaction
HDS	Hampden Data Services
Heilbron&'s	Dictionary of Organic Compounds
HMSO	Her Majesty&'s Stationery Office
HODOC	Handbook of Data on Organic Compounds
HORD	Hierarchically Ordered Ring Description
HOSE	Hierarchically Ordered Spherical Description of Environment
HSDB	Hazardous Substances Data Bank
HTSS	Hierarchical Tree Substructure Search
IAEA	International Atomic Energy Agency
IARC	International Agency for Research on Cancer
ICC	Inter-Company Comparisons
ICP	Index to Conference Proceedings Received
ICR	Intelligent optical character recognition
ICSD	Inorganic Crystal Structure Database

ICSTI	International Council of Scientific and Technical Information
ICSUAB	International Council of Scientific Unions Abstracting Board
IDC	Internationale Dokumentationsgesellschaft für Chemie
IDIOTS	Infrared Spectra Documentation and Interpretation Operating with Transcripts and Structures
IFI	Information for Industry
IFIS	Industry File Index System
IFLA	International Federation of Library Associations
IGOR	Interactive Generation of Organic Reactions
IIA	Information Industry Association
IIB	Institut International des Brevets
IKBS	Intelligent knowledge-based systems
IMF	International Monetary Fund
INID	Internationally Agreed Numbers for the Identification of Data
INIS	International Nuclear Information System
INPADOC	International Patent Documentation Center
INPI	Institut National de la Propriété Industrielle
INSPEC	Information Service for the Physics and Engineering Communities
IPC	International Patent Classification
IPSS	International Packet Switching Service
IRDC	IR Data Committee of Japan
IRSS	Infrared Search System
ISDN	Integrated Services Digital Network
ISHOW	Information System for Hazardous Organics in Water
ISI	Institute for Scientific Information
ISO	International Standards Organization
ISTP	Index to Scientific and Technical Proceedings
ITC	International Translation Center
IUPAC	International Union of Pure and Applied Chemistry
JAICI	Japan Association for International Chemical Information
JANAF	Joint Army, Navy, and Air Force
JCAMP	Joint Committee on Atomic and Molecular Physical Data
JCPDS	Joint Committee on Powder Diffraction Standards
JICST	Japan Information Center of Science and Technology
JIPID	Japanese International Protein Information Databank
JPO	Japanese Patent Office
LAN	Local area network
LHASA	Logic and Heuristics Applied to Synthetic Analysis
MACCS	Molecular ACCess System
MCS	Maximal common substructures
MDL	Molecular Design Ltd.
MEAL	Media Expenditure Analysis Ltd.
MIMD	Multiple instruction stream, multiple data stream
MIPS	Martinsried Institute for Protein Sequences
MISD	Multiple instruction stream, single data stream
MMI	Man - machine interaction
Modem	Modulator — Demodulator
MSDC	Mass Spectrometry Data Center
MSDS	Material safety data sheet
MSI	Marketing Surveys Index
MSSS	Mass Spectral Search System
NASA	National Aeronautics and Space Administration
NBRF	National Biomedical Research Foundation
NBS	National Bureau of Standards
NFAIS	National Federation of Abstracting and Indexing Services
NFPA	National Fire Protection Hazard Rating
NIOSH	National Institute for Occupational Safety and Health
NIST	National Institute of Standards and Technology
NLM	National Library of Medicine
NLP	Natural Language Processing
NMRLIT	Literature Search System
NTIS	National Technical Information Service
OAPI	Organisation Africaine de la Propriété Intellectuelle
OCR	Optical character recognition
ODA	Office Document Architecture
ODIF	Office Document Interchange Format
OECD	Organization for Economic Cooperation and Development
OHM/TADS	Oil and Hazardous Materials/Technical Assistance Data System
OLPI	International Association of Producers and Users of Online Patent Information

OPAC	Online public access catalog
ORAC	Organic Reactions Accessed by Computer
ORTEP	Oak Ridge thermal ellipsoid plot
OSAC	Organic Structures Accessed by Computer
OSI	Open System Interconnection
PAIRS	Program for the Analysis of Infrared Spectra
PATDPA	Patente des Deutschen Patentamts
PATOLIS	Patent Online Information System
PATOS	Patent-Online-System
PBM	Probability-based matching
PC	Personal computer
PCT	Patent Cooperation Treaty
PDG	Patent Documentation Group
PIR	Protein Identification Resource
POSSUM	Protein Online Substructure Searching - Ullman Method
PPDS	Physical Property Data Service
PROLOG	PROgramming in LOGic
PSS	Packet Switching Service
PSTN	Public Switched Telephone Network
QI	Quality Index
QSAR	Quantitative structure – activity relationship
RAM	Random access memory
RAPRA	Rubber and Plastics Research Association
REACCS	REaction ACCess System
ReSy	Research System
RISC	Reduced instruction set computer
ROM	Read only memory
ROSDAL	Representation of Structure Diagram Arranged Linearly
RTECS	Registry of Toxic Effects of Chemical Substances
S4	Softron Substructure Search System
SAHO	Spectral Appearance in Hierarchical Order
SANDRA	Structure and Reference Analyzer
SANSS	Structure and Nomenclature Search System
SCORE	Scan Conversion for Outline Representation of Images
SDI	Selective dissemination of information
SECS	Simulation and Evaluation of Chemical Syntheses
SEMA	Stereochemically Extended Morgan Algorithm
SGML	Standard Generalized Mark-up Language
SIGLE	System for Information on Gray Literature
SIMD	Single instruction stream, multiple data stream
SISCOM	Search for Identical and Similar Components
SISD	Single instruction stream, single data stream
SMD	Standard Molecular Data
SMILES	Simplified Molecular Input Line Entry System
SOLUB	Aqueous Solubility Data Base
SQL	Standard Query Language
SRD	Standard Reference Data
SSI	Small scale integration
STIRS	Self-Training, Interpretive, and Retrieval System
STN	Scientific and Technical Information Network
TDRS	Text Data Retrieval System
THOR	Thesaurus-Oriented Retrieval
TIB	Technische Informationsbibliothek
TLV	Threshold Limit Value
TORC	To Ring-Code
TOSCA	Toxic Substances Control Act
TRC	Thermodynamics Research Center
TSCA	Toxic Substances Control Act
TSCATS	Toxic Substances Control Act Test Submissions
ULSI	Ultralarge scale integration
UMI	University Microfilms International
USPTO	United States Patent and Trademark Office
VCI	Verband der Chemischen Industrie
VDU	Visual display unit
VINITI	All-Union Institute for Scientific and Technical Information
VLSI	Very large-scale integration
WDC	World Data Centers
WHO	World Health Organization
WIMPs	Windows, icons, mice, and pointing devices

WIPO	World Intellectual Property Organization
WLN	Wiswesser Line Notation
WMSSS	Wiley Mass Spectral Search System
WORM	Write-once - Read many (optical disc)
WPA	World Patent Abstracts
WPI	World Patents Index
WPIM	World Patents Index Markush
WYSIWYG	What You See is What You Get

Table of Contents

1   The Scientific Journal

1.1   The Primary Literature

The primary literature includes journal articles, patents, theses, reports, and conference papers. Patents are considered in Part 2. Theses, reports, conference papers, and related items belong to what is often known as the "'gray" literature (Chap. 4). The present chapter concerns original published papers in scientific journals.

To read the primary literature the scientist needs access to a library. WOLMAN has written a useful introduction to library organization for the user of chemical information [1.1]. There are several standard works on chemical information sources [1.2]–[1.8].

A number of specialized journals (e.g., *Scientometrics* and the *Journal of Documentation*) are devoted to studies of the primary and secondary literature. Books are also available about scientific publishing [1.9], [1.10]. The Primary Communications Research Center at Leicester (UK) was a center of expertise from 1976 until 1986 [1.11].

1.2   Trends in Scientific Publications

The first scientific journal, *Philosophical Transactions of the Royal Society*, was published in 1665. At the beginning of the nineteenth century there were about 100 scientific journals. According to Ulrich's International Periodicals Directory, 10 000 serials titles were published in 1951 and 71 000 in 1987. The highest growth rate occurred in the 1960s and was particularly evident for the sciences. Growth in most fields slowed in the 1980s. Chemical Abstracts now handles about 10 000 journals in chemistry or chemically related disciplines. The increase in the number of scientific papers included in Chemical Abstracts between 1907 and 1986 is shown in Figure 1.

As the scope and volume of chemistry have increased, some journals have had to divide into new sections and many new specialized journals have appeared. The exponential growth in journal size is leveling off.

Scientific journals suffer from ever-increasing production costs and, consequently, higher subscription rates. The increasing number of journals and their increased subscription costs cause libraries to give serious consideration to which subscriptions they should discontinue.

There has also been a noticeable trend in language of publication. Germany was the leading scientific nation up to the beginning of the twentieth century but since

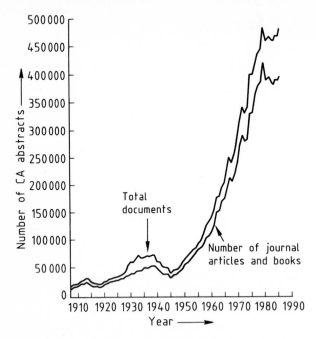

Figure 1. Number of scientific papers included in Chemical Abstracts from 1907 to 1986 (reproduced with permission from [1.12])

then the United States has achieved dominance of the scientific literature. Table 1 shows the rise in the use of English since 1961.

Table 1. Language of publication of journal literature abstracted in Chemical Abstracts, as percentage of total journal literature abstracted

Language	1961	1966	1972	1978	1983	1988
English	43.4	54.9	58.0	62.8	68.6	73.4
Russian	18.4	21.0	22.4	20.4	15.8	12.0
Japanese	6.3	3.1	3.9	4.7	4.4	4.1
German	12.3	7.1	5.5	5.0	3.5	3.3
Chinese	*	0.5	*	0.3	1.9	2.8
French	5.2	5.2	3.9	2.4	1.5	1.1
Polish	1.9	1.8	1.2	1.1	0.8	0.7
Spanish	0.6	0.5	0.6	0.7	0.5	0.5
Czech	1.9	0.9	0.6	0.5	0.4	0.3
Italian	2.4	2.1	0.8	0.6	0.5	0.3
Korean	*	*	0.2	0.2	0.3	0.3
Others	7.7	2.9	2.9	1.3	1.8	1.2

* Included in "Others" for year.

1.3   Function of the Scientific Journal

The journal serves as a foundation for the advancement of science. Its main functions are archival storage, current awareness, quality control, and author recognition [1.13], [1.14].

Until the mid-nineteenth century, journals were archival in character, recording the text of papers presented orally at scientific meetings. Scientists started to rely more upon journals rather than meetings for their information and journals began to develop an alerting or current awareness function.

Nowadays, journal scanning is still useful for current awareness but probably not as useful as "the invisible college"—conferences, seminars, informal visits, letters, and telephone calls. Formal papers are not written until the completion of a project, and publication and abstracting delays diminish the current awareness value of the published article. Publication delays of up to 8 months or more are inevitable because of the quality control process. Authors submit papers and the editor sends them to referees for assessment. Quality control is discussed further in Section 1.4.

1.4   Quality and Prestige

The range of journals in circulation provides a subtle system of quality control. A reviewer may reject an article for one journal but accept it for a "lesser" journal. Not all journals are "learned"; there is a place also for "news" journals such as *Chemical and Engineering News* or *Chemistry in Britain*.

Articles in learned journals are of four types: full paper, note, communication, or review. A note is shorter than a full paper and describes more limited findings. A communication is more urgent and provides a preliminary report of important results. Reviews are considered in Section 3.2.

Broadly speaking, journals with a high rejection rate (by the editor and/or the peer reviewers) have high prestige. Interdisciplinary primary journals such as *Science*, *Nature*, or *Experientia* are likely to have a particularly high rejection rate: over 90% for *Science* compared with about 45% for the *Journal of Chemical Information and Computer Sciences*.

Due to increasing costs and number of publications, librarians use various criteria for selection of journals, for example citation analysis of the primary and secondary literature, and journal-use pattern studies [1.15].

The theory of *citation analysis of the primary literature* [1.16]–[1.19] is as follows. When researchers cite a journal article, it indicates that this article has influenced them. The more frequently that a journal is cited, the more often the scientific community indicates the influence or impact of that journal. Science Citation Index (see Section 2.3.1) has a data base of literature citations and it produces an annual analysis called the *Journal of Citation Reports* [1.20]. In 1987 eight chemical journals were listed amongst the 30 most cited journals:

Journal of Biological Chemistry
Journal of the American Chemical Society
Journal of Chemical Physics
Biochimica et Biophysica Acta
Biochemistry
Biochemical Journal
Journal of Organic Chemistry
Journal of Physical Chemistry.

Dangers in rating the prestige or quality of a journal purely on its rank in such a list are discussed in [1.19]. New journals, journals that subdivide, and journals that change title are at a disadvantage in citation analysis ranking. Some important journals are often read but little cited. Journal reputation, circulation, availability, and the extent of library holdings and coverage by secondary services, all have an impact on the citation frequency.

Ranked lists of journals can also be produced according to the frequency of their *citations in the secondary literature*. The Chemical Abstracts Service Source Index [1.21] lists the most frequently cited journals in Chemical Abstracts. As in the Science Citation list, the *Journal of Biological Chemistry* is first (1988), *Biochimica et Biophysica Acta* is second, and the *Journal of the American Chemical Society* fourth. Such rankings have certain merits compared with those produced from citation analysis of the primary literature. Recently published journals have a better chance of finding a higher place as do popular journals, such as *New Scientist*.

Many of the most prestigious journals are produced by learned societies but there are also several highly regarded commercial publishers.

There is an unproven theory that higher prestige in scientific journals goes hand in hand with low readability [1.22]. There are many formulae for measuring readability [1.23], [1.24].

1.5   Document Delivery

If the article a scientist wants is not available in the library, he may request a photocopy from another library, use a commercial service to obtain a copy, or write to one of the article's authors to obtain a reprint. Many scientists have access to a library that can arrange an interlibrary loan.

The developing nations [1.25], [1.26] and the Eastern Bloc [1.27] have particular problems because their libraries do not stock many journal titles and obtaining copies of articles may be costly and difficult. Writing to an author for a reprint is the cheapest option.

Many guides or periodicals enable users to locate primary journals and check titles and publication data. In North America the first sources to consult are the Union List of Serials in Libraries of the United States and Canada, or Chemical Abstracts Service Source Index [1.21]. Other countries have their own national sources (e.g., the British Library).

The use of optical storage media, electronic mail, and telefacsimile opens the way for the development of large-scale automated document delivery systems [1.28]–[1.31]. See also Section 8.2.10.

Libraries and information centers which operate document supply services include:

The British Library Document Supply Center
Colorado Technical Reference Center
Delft University of Technology Library
Technical Research Center of Finland
Centre National de la Recherche Scientifique (France)
Technische Informationsbibliothek (Germany)

The cost of these services is beyond the reach of most developing countries. A cooperative solution has been proposed for the International Information System for the Agricultural Sciences and Technology, AGRIS [1.26].

Commercial services for document delivery are operated by Chemical Abstracts Service (their Document Delivery Services, CASDDS), by the Institute for Scientific Information (The Genuine Article) and by the University Microfilms International Article Clearing House. Some private organizations also offer services (e.g., INFO from Information on Demand Inc. of Berkeley, California).

1.6   Translation

English is the dominant language of scientific publishing (see Table 1) and many publications appear totally or partially in English even though that is not the native language of the publisher's country. *Chemische Berichte* and *Angewandte Chemie* are notable examples. Nevertheless large numbers of articles are published in other languages, and scientists may require English translations. Non-English speaking scientists also require translations of articles published in English.

The scientist should first try to find an abstract either in the original journal (many journals produce abstracts in more than one language) or from an abstracting service (see Chap. 2). If the abstract looks interesting, a translation of the whole article may be needed.

Many journals, especially those in Russian, are translated into English cover to cover. The Chemical Abstracts Service Source Index [1.21] and the indexes produced by national translation centers are convenient ways of identifying and locating such journals.

In most developed countries translation centers collect translations into their native language, e.g., the National Translation Center at the John Crerar Library in Chicago, USA and the Centre National de la Recherche Scientifique, CNRS, France. The International Translation Center (ITC) in Delft, The Netherlands, is a translation clearing house which was originally set up to translate "difficult" languages into Western European ones but which now also handles translations of European languages. It prints the World Index of Scientific Translation. World Translations Index, a file produced by CNRS and ITC, is available online on DIALOG.

If a translation of an article is not available, a near equivalent may possibly be found, for example an article in another language by the same author, a review article, or a book. If, however, a ready-made translation or an equivalent cannot be found, the scientist will have to use a dictionary [1.1], ask a colleague to do a translation, or pay for a custom-made translation. The use of computers in translating is considered in Section 11.5.2.

1.7   Copyright

In many countries "writings" are protected by copyright law e.g., the Copyright, Designs and Patents Act 1988 in the United Kingdom. In the United States the Copyright Law enacted in October 1976 became effective on January 1, 1978 and will be described here as an example.

For works created on or after January 1, 1978, copyright lasts for the life of the author plus 50 years.

Copyright is not restricted to the written word. Composers, artists, and computer programmers are also protected. A "writing", unlike a patentable invention, need not be novel. The copyright owner has the exclusive right to reproduce the work, prepare derivative works, distribute copies and, in the case of musical and dramatic works and the like, to perform and display the work publicly.

There are "fair-use" exceptions to these exclusive rights. For example, copies may be taken for teaching, scholarship, and research provided certain rules are obeyed. Libraries and archives can make and distribute a single copy of any work if they do so without commercial advantage and conform to certain restrictions.

The law also covers copyright transfer. Nowadays most publishers of scientific papers require copyright transfer before publication. They can then more easily grant permissions to reproduce, make text available in full text form for online searching, and so on.

Useful summaries of the law and its impact have been published [1.32]–[1.34]. The Copyright Clearance Center operates a scheme of collection of authorization fees for photocopying copyright publications [1.35].

Copyright statute revision has always lagged behind technology developments. New guidelines are needed to protect the rights of creators and copyright owners, while allowing scientific users to harness the advantages of new technologies in electronic document replication and delivery.

1.8   Alternatives to the Conventional Journal

A tendency to conservatism exists in the domain of primary communication. The price of inappropriate innovation is high for publishers and editors. Authors wish to perpetuate the system on which their prestige depends. The majority of journals are still, therefore, printed on paper, but other methods of information transfer have been investigated [1.36].

Conventional journals can be reproduced in microform but most such journals have corresponding printed editions. Microform versions can be useful for libraries short of shelf space.

For many years there have been discussions about the value of synopsis or abstract journals [1.37], [1.38]. The current awareness function is supposed to be served by the synopsis while the archiving function is served by the filing of microfiche or separates. The *Journal of Chemical Research* appears as a synopsis journal plus a microfiche and a miniprint version.

The production of supplementary material to published journals is much more common; experimental detail, crystallographic structure data, and computer programs, are deposited separately in a repository. The United States has had a national depository (the National Auxiliary Publications Service) since 1937. The British Library Document Supply Center has had a similar scheme since 1969. In the Soviet Union, VINITI (the All-Union Institute for Scientific and Technical Information) runs a deposition scheme. The Royal Society of Chemistry and the American Chemical Society also store supplementary material for certain journals.

Computers have also been introduced into publishing [1.39]. "Electronic publishing" refers to the use of the computer in the production of publications and also in the electronic distribution of text via computer terminals. The use of a generalized mark-up format for electronic publishing of a hard-copy journal can facilitate conversion of data into an online full-text data base [1.40]. The Chemical Journals Online (CJO) service on Scientific and Technical Information Network (STN) International contains the full text of the American Chemical Society's primary research journals (CJACS), the Royal Society of Chemistry's ten primary journals (CJRSC), J. Wiley and Son's five primary polymer journals (CJWILEY), the Association of Official Analytical Chemists' primary journal (CJAOAC), Elsevier Journals (CJEL-SEVIER) and VCH's international edition of *Angewandte Chemie* (CJVCH) [1.41]. Full-text data bases are considered in Section 8.2.8.

Pergamon's *Tetrahedron Computer Methodology* claims to be the first scientific journal to be simultaneously published in hard copy and in electronic form (floppy disks).

2 Abstracting and Indexing Services

2.1 Introduction

Scholars have always been concerned with the speedy identification of, location of, and access to the contents of learned journals. The need for devices to aid them increased with the explosive growth of the primary literature [2.1]. Abstracting and indexing services (sometimes known as secondary information services, or the secondary literature) are designed to help scholars keep abreast of the primary literature. It is estimated that more than 1500 abstracting–indexing publications worldwide are concerned with learned publication.

The major abstracts journals in chemistry, many of which are now discontinued, are listed in Table 2. (Patent services are considered in Chaps. 20 and 21)

Various common-interest groups in secondary information have emerged in the United States including the International Council of Scientific and Technical Information (ICSTI), formerly (before 1984) the International Council of Scientific Unions Abstracting Board (ICSUAB), the National Federation of Abstracting and Indexing Services (NFAIS), the Information Industry Association (IIA), and the Association of Information and Dissemination Centers (ASIDIC) [2.2]. In Europe, the European Association of Information Services (EUSIDIC) and the European Information Industry Association (EIIA) have more generalized interests than abstracting and indexing. A Global Alliance of Information Industry Associations (GAIIA) was set up in 1989.

Abstracts originally served an alerting function: they gave scholars a quick and easy method of keeping up with the rapidly growing literature. Abstracts are now an indispensable filtering tool in retrospective access [2.3]. The scientist who has located

Table 2. Abstract journals in chemistry

Journal	Year published
Chemisches Zentralblatt	1830–1969
Chemical Abstracts	1907–present
British Abstracts	1926–1953
Nippon Kagaku Soran, Second Series*	1927–1974
Bulletin Signalétique	1940–1983
Referativnyi Zhurnal, Khimiya	1953–present
Current Abstracts of Chemistry and Index Chemicus	1960–present
Chemischer Informationsdienst (ChemInform)	1970–present
Kagaku Gijutsu Bunken Sokuho, Kagaku, Kagakukogyo Hen*	1974–present

* Nippon Kagaku Soran changed to Kagaku Gijutsu Bunken Sokuho, Kagaku, Kagakukogyo Hen (Current Bibliography on Science and Technology, Chemistry and Chemical Engineering) in 1974.

an article in a printed index, or from an online search, can scan an "informative" abstract to decide whether it is worth reading the original article in full. An informative abstract is defined by the American National Standards Institute (ANSI) as follows [2.4]:

"A well-prepared abstract enables readers to identify the basic content of a document quickly and accurately, to determine its relevance to their interests, and thus to decide whether they need to read the document in its entirety."

2.2 Chemical Abstracts Service [2.5], [2.6]

The American Chemical Society started to publish *Chemical Abstracts* (CA) in 1907, to make the results of chemical research throughout the world accessible to scientists in the United States. At this time more publications appeared in German than in English, and American chemists were dissatisfied with coverage of the American literature in European abstracting journals. What began as a national organization has become the international *Chemical Abstracts Service* (CAS) based in Columbus, Ohio. The American Chemical Society gave CAS the mission of abstracting the complete world chemistry literature. The history of abstracting at CAS is bound up with the definitions of "complete", "abstract," and "chemistry" [2.7]. The motto of Editorial Operations at CAS today is "quality, comprehensiveness, and timeliness".

CAS employs nearly 1500 people, over a third of them chemists. Highly trained staff analyze and input information, and computers assist in editing and reviewing prior to publication.

Over 1.5×10^6 documents are examined every year in the fields of chemistry, chemical engineering, and related sciences.

The median currency ("up-to-dateness") of all abstracts in CA, measured from the publication date of the original article to the date the abstract was published in CA, is about 90 days. For the nearly 800 journals covered by Chemical Titles (see Section 2.2.5), the median is less than 2 months.

2.2.1 Statistics

Just under 12 000 abstracts were published by CA in 1907; in 1989, nearly 490 000 were published and the cumulative number of abstracts had risen to nearly 12.75×10^6. It took CA 30 years to publish the first 1×10^6 abstracts. Nowadays nearly 1×10^6 abstracts are added every two years. Over 1.5×10^6 documents are examined by CAS each year, including patent documents, conference proceedings, government reports, books, and 10 000 scientific journals. The documents come from more than 150 countries and are written in more than 50 languages.

By the end of 1989 nearly 10×10^6 chemical substances were known to the CAS Registry System (see Sections 2.2.2 and 2.2.5); over 600 000 new compounds were registered in 1989.

The eleventh collective index (see Section 2.2.3) is reputedly the world's largest index with 28×10^6 entries.

2.2.2   Document Analysis

In the early days CAS used large numbers of volunteer abstractors. Nowadays more than 95% of the abstracts are prepared in-house. The Japan Association for International Chemical Information (JAICI) abstracted and indexed over 13 000 Japanese patent documents for CAS in 1988 and the Fachinformationszentrum Chemie (FIZ Chemie) in Germany abstracted and indexed more than 4000 German-language documents. CAS document and structure analysts have expertise in a wide range of scientific disciplines and most of them have skills in languages other than English.

A document analyst starts by selecting items from the document and translating titles into English. Editorial Coordination Services flag key bibliographic information for data entry. These data are entered and then checked by a second keyboard operator. Each source document and an abstract preparation sheet are then sent to a document analyst, who prepares an informative English-language abstract and generates index entries and keywords. Documents are batched up and converted into computer-readable form.

At this point it is necessary to index, or "register" chemical structures into the CAS Registry System. CAS assigns a Registry Number of the form [*nnnnnn-nn-n*] (where *n* is a one-digit number) to each unique compound. In order to check whether a compound is novel its name is first matched against the Registry Nomenclature File. If no name-match is found, either the name is a new synonym for a known compound (and will need adding to the Registry Nomenclature File) or the compound is novel. A registry structure sheet is generated for compounds that are not successfully name-matched or that have multiple index names. These structure sheets are sent to an analyst who resolves the name-match hits and draws structures for new compounds.

Another analyst then edits the abstract and index work units online. Batches of data without structure sheets are released for publication; structure sheets are sent on to the structure input process.

Each structure diagram is converted into computer-readable form. The CAS Chemical Registry System converts the machine structure record into a unique connection table by means of the Morgan algorithm [2.8]. The connection table (discussed further in Section 10.1.4) can be matched against the 10×10^6 other substances in the structure file. If a match is found, the previously assigned Registry Number is retrieved. If no match is found, a new Registry Number is automatically assigned by the computer and its record is added to the file.

Registry editors then resolve inconsistencies before new structure registration sheets are sent on to nomenclature specialists. Staff assign unique and unambiguous index names to each new substance (see Section 2.2.4). These names are keyed into the computer and added to the CAS ONLINE Registry File.

At the next stage, the computer selects appropriate material for a particular publication or service and converts it into the proper format. A final quality check is made before pages of CA and its indexes are dispatched for printing and distribution. Other CAS services are produced in-house. The data from CA and its indexes are also added to STN International for online searching.

2.2.3   Indexes

Documents are abstracted and indexed according to five main subject areas and 80 sections. CA indexes were published annually until 1962 and thereafter semiannually. The CA Collective Indexes combine the contents of ten individual volume indexes into single, organized listings [2.6], [2.9]. The Eleventh Collective Index contains author, chemical substance, general subject, patent, and formula indexes; an index of ring systems; and an Index Guide. The Index Guide explains the use of the indexes and gives an extensive summary of CA nomenclature.

2.2.4   Chemical Nomenclature [2.10]–[2.12]

There are many different types of nomenclature, each having advantages and disadvantages for certain purposes. CA nomenclature is designed for use in a large printed index where each name must be unique and unambiguous [2.13].

The International Union of Pure and Applied Chemistry (IUPAC) is responsible for an international standard in chemical nomenclature [2.14]–[2.17]. One substance may, however, have several IUPAC names. CA names conform with IUPAC principles but CA has had to make additional rules to achieve uniqueness.

With the publication of the Ninth Collective Index, CA simplified the nomenclature system and made it more systematic [2.18]. The systematic nomenclature has, however, some disadvantages: names tend to be longer and relatively unfamiliar names are used for familiar compounds, e.g., benzenamine for aniline.

Fully systematic nomenclature should allow a computer to generate chemical names for chemical structures or vice versa. The introduction of computers to the CAS system in the early 1960s led to the development of well-defined translation procedures for converting CAS systematic names to atom–bond connection tables [2.19]. Computer programs were used to validate the CAS index names in early records and to convert them to a common internal form for storage [2.20]. These routines have since been extended to provide editing and verification facilities for new index names [2.21].

The University of Hull (UK) has developed microcomputer software to convert chemical names into connection tables, from which chemical structures can subsequently be displayed [2.22]–[2.24].

In 1984–1985 the Beilstein Institute developed a program, VICA, to convert German chemical names into structures on a mainframe IBM computer for the first version of the Beilstein Online data base (see Section 9.6.1) [2.25]. The program is being developed to handle English names. The Beilstein Institute is also developing a microcomputer program, AUTONOM, which will convert structures to names [2.25]. At the moment neither VICA nor AUTONOM handles stereochemistry.

2.2.5   Computerization at CAS

The world's first computer-produced periodical, Chemical Titles, was introduced by CAS in 1961 and covers ca. 750 chemically oriented journals. Each article is indexed by author and keyword with a reference to the journal in which it appears and each is subsequently reported in CA.

The bibliographic CA data base has been compiled since 1967 and is accessible for online searches (see Section 8.2.4).

Since 1965 all chemical substances mentioned in the literature have been recorded according to the CA Registry System [2.26]–[2.36]. Pre-1965 data are now being gradually added. A two-dimensional structural formula, stereochemical descriptor, all labeled atoms, and unusual valences are recorded for each registry number. In addition, the molecular formula, the CA Index Name, and all known trivial and trade names are linked to the CA Registry Number.

The Registry File can be used to ascertain whether a substance with a certain name has already been mentioned under the same or another name. About 6000 substances are checked by CA every day.

In the mid-1980s CAS began a Registry Enhancements project to determine how the Registry System might be improved in response to user demand. Important aspects of the project are changes in the areas of polymers, biological macromolecules, stereochemistry, coordination compounds, and inorganic compounds and materials [2.37].

2.2.6   Other CAS Products and Services

The Ring Systems Handbook [2.6] which first appeared in 1984, contains information about all ring and cage systems in the CAS Chemical Registry System. The Registry Handbook–Common Names on microform lists synonymous common names and CA Registry Numbers. CAS run a search service on the CAS data base and on a biosequence data base on request.

Chemical Industry Notes, both computer-readable and in hard copy, is a weekly journal of extracts from about 100 leading industry and trade journals. The extracts are grouped into eight sections: production, pricing, sales, facilities, products and processes, corporate activities, government activities, and people.

To provide convenient international access to data bases in a wide range of scientific and technical areas, CAS has joined with FIZ Karlsruhe in the Federal Republic of Germany, and the Japan Information Center of Science and Technology (JICST), to operate the Scientific and Technical International Network known as STN International. Some of the online data bases on this service are considered in Chapter 8. CAS also support microcomputer software to help users access some of these data bases (see Section 10.6.4). CAS and FIZ Karlsruhe are also cooperating in the development of a numeric data service. Some numeric data bases are already available on STN International (see Sections 9.4.1, 9.6.1, and 9.6.4).

CAS provide support for registration of substances in, and addition of CAS Registry Numbers to, a number of non-CAS data bases, for example BioSciences Information Service's BIOSIS Previews File. CAS have also applied their expertise in software development to the automation of patent processing in the U.S. Patent and Trademark Office.

2.3   Institute for Scientific Information

The Institute for Scientific Information (ISI) is a for-profit organization, founded in 1960. It has about 650 employees worldwide but is based in Philadelphia. ISI is best known in the field of citation indexing but it also produces other important abstracting and indexing publications and services. ISI's current awareness products are covered in Section 2.5.

2.3.1   Science Citation Index

An article in a learned journal traditionally cites references to other related articles. For the purpose of citation indexing, the references are *cited articles* and the article citing them is the *citing article*. A *citation index* is an ordered list of cited articles each of which is accompanied by a list of citing articles. The citing article is identified by a source citation, the cited article by a reference citation. The index is arranged by reference citations. Any source citation may subsequently become a reference citation.

Use of a citation index enables a searcher to locate a chain or network of related articles, starting from one literature reference, without the need to use keywords, subject terms, or chemical names. Scientists can determine which subsequent papers have cited a particular reference and can thus move both backwards and forwards in time in a literature search.

Science Citation Index [2.38], [2.39] was first published in 1961; it is now published six times a year and cumulated annually. There are also five- and ten-year cumulations. The Citation Index is a series of indexes and is now available in printed form, as an online data base, and as a compact disc (see Chap. 8). It covers annually about 3500 journals and a few hundred nonjournal titles from more than 100 disciplines, including chemistry. The Citation Index lists cited documents alphabetically by name of the first author. The Source Index is alphabetically arranged by author with a separate index by organization. The Permuterm Subject Index is based on the original words in the titles of items covered. All significant words in a title are coupled together.

The theory of citation analysis is briefly discussed in Section 1.4. Another extension of citation indexing, cocitation clustering, is now also used for automatic hierarchical classification and mapping of the literature [2.39].

2.3.2   Index Chemicus

Index Chemicus (between 1970 and 1986 called Current Abstracts of Chemistry and Index Chemicus) is a weekly abstracting and indexing service which has been published since 1960. It details new organic compounds and syntheses reported in about 110 of the journals most important to organic chemists. It is claimed that over 90 % of all new organic compounds appear in this small selection of journals. Index Chemicus endeavors to print an abstract less than 45 days after publication of the original article. Unfortunately patents are not covered. The strong points of Index Chemicus are currency and the easy-to-read, highly graphic abstracts that include structural diagrams and reaction flows. The ISI accession number permits ordering of the original article through ISI's The Genuine Article service (see Section 1.5). The indexes to Index Chemicus are cumulated quarterly and annually and include a permuted (rotated) molecular formula index.

2.3.3   Current Chemical Reactions

This ISI tool was first published in 1979. It prints flow diagrams of new and newly modified reactions and syntheses, with abstracts and bibliographic details, abstracted from over 120 journals and some books. About 300 000 reactions are reported annually.

Current Chemical Reactions is also available as an in-house computer-readable data base for use with Molecular Design Limited's REACCS software (see Section 10.8).

2.4   Other Abstracting and Indexing Services

The abstract journals listed in Table 2 (p. 8), even those now discontinued, are useful in addition to CA because they cover time periods CAS does not and they have a different scope.

ChemInform (formerly Chemischer Informationsdienst) has been produced since 1970 by FIZ Chemie and Bayer AG. It is an abstracting journal that mainly covers publications on new reactions and synthetic methods from about 250 major journals. It is published weekly, since 1987 fully in English.

Online data bases produced by abstracting and indexing services are described in Chapter 8. For further details of the hard-copy (and online) products, see [2.6], [2.10], [2.40]–[2.45]. A large directory of abstracting and indexing services is available [2.46]. Important services are listed below:

The Royal Society of Chemistry produces a number of current awareness bulletins (Section 2.5).

In the field of agricultural science CAB International, previously called the Commonwealth Agricultural Bureaux, provides an outstanding abstracting and indexing service.

The National Library of Medicine in the United States and Excerpta Medica in Europe are important sources of medicinal chemical information.

The Rubber and Plastics Research Association (RAPRA) produces abstracts, indexes, and an online data base.

BioSciences Information Service (BIOSIS) produces Biological Abstracts and other services.

Engineering Information produces Engineering Index and its computer-readable form COMPENDEX.

Information Service for the Physics and Engineering Communities (INSPEC) is a product of the Institution of Electrical Engineers in the United Kingdom.

The Deutsche Gesellschaft für chemisches Apparatewesen, Chemische Technik und Biotechnologie (DECHEMA) abstracts and indexes about 400 engineering journals.

Cambridge Scientific Abstracts are a large for-profit publisher of abstracts in several fields.

Chemical reactions are covered by Current Chemical Reactions (see Section 2.3.3), Methods in Organic Synthesis from the Royal Society of Chemistry (see Section 2.5), and the compendia described in Section 3.4.

2.5   Current Awareness Services

Many of the vendors mentioned earlier in this chapter also produce current awareness services. Indeed some of the hard-copy products such as Current Chemical Reactions are used by scientists for current awareness.

Selective dissemination of information (SDI) refers to those activities that help researchers stay up to date with the literature on a specific topic. A search strategy used for an online data base search can be saved and then run periodically against updated parts of the data base. Most search systems can also issue a series of commands which automatically run the strategy against each new update of one or more data bases and have printouts made on a regular basis. Such a stored strategy is known as an *SDI profile*.

The Royal Society of Chemistry produces several current awareness periodicals and data bases:

Chemical Hazards in Industry
Laboratory Hazards Bulletin
Methods in Organic Synthesis

Natural Product Updates
Current Biotechnology
Process and Chemical Engineering
Theoretical Chemical Engineering Abstracts
Mass Spectrometry Bulletin
Chemical Business Bulletins
Chemical Business Update

The ISI weekly service, Current Contents, reproduces tables of contents of about 1000 of the most important research journals in chemistry and related sciences. A similar publication, Chemical Titles, from CAS, is described in Section 2.2.5.

Automatic Subject Citation Alert (ASCA) is an SDI service based on the ISI online data base. ASCATOPICS (renamed Research Alert, as of Feb. 1st 1990) is a series of about 350 profiles, any selection of which can be run by ISI as a service for a scientist. Patent information is, however, not included and there are no abstracts.

CA SELECTS is a very valuable current awareness service supplied by CAS. It provides complete CA abstracts and bibliographic citations, and also covers patents but is not as fast as ASCATOPICS. The number of topics in 1989 was 209.

2.6   Future of Abstracting and Indexing Services

The continuing growth of the primary literature has led to continuing expansion of secondary services. There has been a "migration" from print products to electronic services and this is likely to continue as more countries and users gain access to online data bases.

A trend towards in-house systems, not simply CD-ROM data bases (Chap. 8), is likely. In-house systems are more cost effective for regularly used data, there are fewer problems with telecommunications, and the software can be tailored better to the users.

More and more "end users" are gaining access to online information, bypassing the services of an intermediary information scientist. The intermediary is, however, usually needed for complex searches.

New technology (Chaps. 6 and 7) will continue to have an impact on information services. One result is the integration of primary and secondary publishing. When journals are photocomposed, the machine-readable text can be reused in other ways, including input to secondary services (see Section 1.8). As the full text of journals becomes available online the traditional boundaries between primary and secondary publishing will blur.

3   Tertiary Literature

3.1   Introduction

Definitions of tertiary literature vary but here it is assumed to be evaluated literature based on primary and secondary sources, including reviews, handbooks, and encyclopedias. For convenience, other standard reference works in chemistry are also included.

Journals and patent documents nearly always contain more current information than books. Encyclopedias, even more so than books, have a problem with obsolescence. Nevertheless good books and encyclopedias are an excellent source for an overview or an evaluation.

This chapter covers a selection of the most important reviews, books and encyclopedias. There are several standard works on chemical information sources that are more comprehensive [3.1]–[3.8].

3.2   Reviews

Because of the enormous growth in chemistry a review writer cannot both cover a topic from its origins and evaluate all recent references. Reviews are therefore more suitable for updating a researcher's knowledge than for providing the educator with an overview.

Key review journals include Chemical Reviews (American Chemical Society), Annual Reports on the Progress of Chemistry (Royal Society of Chemistry), and Chemical Society Reviews (Royal Society of Chemistry). The Society of Chemical Industry produces Critical Reports on Applied Chemistry. Russian Chemical Reviews provide an English translation of Uspekhi Khimii. The Royal Society of Chemistry has produced more than 40 titles in its series Specialist Periodical Reports.

Reviews are designated in the Volume Indexes of Chemical Abstracts by the letter R. The Institute for Scientific Information's Index to Scientific Reviews is available both in hard copy and online.

3.3   Encyclopedias and Handbooks

3.3.1   Encyclopedias

Kirk–Othmer. The Encyclopedia of Chemical Technology [3.9] is commonly referred to as Kirk–Othmer after the names of the original editors. The third edition was produced between 1978 and 1984. It consists of 24 volumes, a supplement, and an index, the main work containing over 1200 articles, 9×10^6 words, 6000 tables, and 5000 figures. Chemical Abstracts Registry Numbers were included.

The articles are well set out and easy to read with emphasis on applied chemistry. Kirk–Othmer is an indispensable tool in almost every chemistry library but has a bias towards American practice. The full text of this encyclopedia is available online and on CD-ROM.

A one-volume version, Kirk–Othmer Concise Encyclopedia of Chemical Technology, was published in 1985. Subject-oriented reprint volumes (e.g., on antibiotics) are also available.

Ullmann's. The fourth edition of Ullmanns Enzyklopädie der Technischen Chemie [3.10], in 25 volumes (in German) began in 1972 and was completed in 1984. The fifth edition, under the title Ullmann's Encyclopedia of Industrial Chemistry, is being published in 36 volumes in English [3.11]. Work began in 1984 and is expected to be complete by 1996. Volumes A1–A28 contain articles about industrial chemicals, product groups, and production processes covering all branches of the chemical and allied industries. The eight B volumes form a basic knowledge series covering chemical engineering fundamentals, analytical methods, environmental protection, and plant safety. Each year a cumulative index appears.

Ullmann's Encyclopedia is international in authorship and coverage. To aid the reader, all but the shortest articles start with a table of contents and the printing is in easy-to-read columns.

The advantages and disadvantages of Ullmanns in comparison to Kirk–Othmer are discussed in [3.12] which, however, erroneously assumes that there is a significant price difference between the two.

3.3.2   The Handbuch Concept

The Handbuch is an old tradition in German chemistry. It is quite different from U.S. handbooks, being multivolume and more extensive in scope and coverage.

In the United States, Handbuch volumes tend to be regarded as obsolete, incomprehensible, and difficult to use. It does take months, or even years, to evaluate data for an accurate and reputable handbook, but coverage often goes back to the beginnings of chemistry, which is not true of, for example, Chemical Abstracts.

At least two major handbooks (Beilstein and Gmelin) are now published in English, making them accessible to a wider scientific audience. New tools are also being produced to facilitate the use of these handbooks.

The Beilstein Handbook. Beilstein's Handbook of Organic Chemistry [3.13]–[3.15] is the oldest, best-known reference work in organic chemistry. It takes its name from Friedrich Konrad Beilstein who produced the first edition between 1881 and 1883. Coverage goes back to the beginning of organic chemistry (1830). Substances are included in Beilstein if they are organic compounds; if they have known, verified constitutions; if they are pure; if syntheses for them are known; and if data are available on them. Information is abstracted from journals, patents, monographs, and other publications. Since 1985 electronic abstracting methods have been employed [3.16]. The Beilstein Institute in Frankfurt, a nonprofit organization, employs over 170 permanent staff, including 120 chemists and 500 external abstractors. The aim is critical evaluation of the primary literature and the ordered presentation of verified facts and data. During evaluation, contradictions are cleared up, duplicate material and trivia are eliminated, and errors in the primary literature are corrected.

Descriptions of compounds cover constitution and configuration; natural occurrence and isolation from natural sources; preparation and manufacture; chemical and physical properties; structural and energy parameters; characterization and analysis; and salts and addition compounds.

Beilstein consists of more than 350 volumes. The original work (basic series) was divided into 27 volume numbers and the five supplementary series adopt the same classification and volume number arrangement (Table 3). The fifth supplementary series is published in English. Up to the end of 1960, 1.5×10^6 compounds had been described. Characteristic of the Handbook is the Beilstein classification system, which is explained in [3.17].

Table 3. The series of the Beilstein Handbook (4th edition)

Series	Abbreviations	Literature years covered	Spine label color	Status
Basic Series	H	Up to 1909	green	complete
Supplementary Series I	EI	1910–1919	dark red (on brown cover)	complete
Supplementary Series II	EII	1920–1929	white	complete
Supplementary Series III	EIII	1930–1949	blue	complete
Supplementary Series III/IV*	EIII/IV	1930–1959	blue/black	complete
Supplementary Series IV	EIV	1950–1959	black	complete
Supplementary Series V** (in English)	EV	1960–1979	red (on blue cover)	1984–

* Volumes 17–27 of Supplementary Series III and IV (heterocyclic compounds) are combined in a joint issue.
** The first volumes published in this series (17–27) relate to heterocyclic compounds based on a survey of user requirements.

Table 4. Contents of Beilstein volumes

Beilstein volume number	Compound class
1−4	acyclics
5−16	isocyclics
17−27	heterocyclics type and number of heteroatoms
17, 18	1 O
19	2 O, 3 O ...
20−22	1 N
23−25	2 N
26	3 N, 4 N ...
27	1 N, 1 O; 1 N, 2 O ...
	2 N, 1 O; 2 N, 2 O ...
	further heteroatoms*

* E.g., B, Si, P but not S, Se, Te.

Organic compounds are distributed over the 27 volume numbers as shown in Table 4. Within these classes further ordering is based on type and number of functional groups. At the top of the even-numbered pages is the designation of the compound class to which the compounds dealt with on that page belong. At the top of the odd-numbered pages are the coordinating reference, the series number, volume number, and system number. The coordinating reference indicates the page of the Basic Series to which the item would have been assigned had the substance been known then. The system number is the unit of the Beilstein System of structure classification. Values run from 1 to 4720 and are dependent on structural features.

If an item is not being covered for the first time, a back reference is given to earlier volumes and page numbers.

To locate a compound in Beilstein the user can use the molecular formula index, the subject (i.e., compound) index, or the system number. Each volume has formula and subject indexes; there are also cumulative and collective indexes. Once the compound has been located in a particular volume, the system number and coordinating reference can be used to find the same compound in the same volume number of other supplements.

A microcomputer program called SANDRA (Structure and Reference Analyzer) aids the user in locating the appropriate volume of Beilstein once he draws in the chemical structure concerned [3.18], [3.19].

Beilstein Online, an electronic version of the Handbook, is now available [3.16], [3.19]. The data base can be searched by structure, substructure, and other features (see Section 9.6.1). The online version contains more compounds than the Handbook because it also contains more current, nonevaluated data.

There are significant differences between Beilstein and Chemical Abstracts [3.19], [3.20] and a well-stocked library needs both. Beilstein covers the literature back to 1830, Chemical Abstracts only to 1907. Chemical Abstracts, however, is much more current than Beilstein. Beilstein gives immediate access to validated numeric data and facts, whereas Chemical Abstracts gives the reader bibliographic references to sources containing such facts. The Beilstein classification locates parents and derivatives on nearby pages. However, Chemical Abstracts permits subject and author searches in the hard copy and Beilstein does not.

Gmelin. The Gmelin Handbook of Inorganic and Organometallic Chemistry [3.21] is an invaluable source of information on elements, inorganic compounds, and organometallic compounds. It is named after LEOPOLD GMELIN, who published his first Handbuch from 1817 to 1819. The handbook is now produced by the Gmelin Institute of Inorganic Chemistry and Related Sciences, which is part of the Max Planck Society for the Advancement of Sciences. The Gmelin Institute is housed in the same building in Frankfurt as the Beilstein Institute and employs almost as many scientific staff.

Work on the current eighth edition was begun in 1922. This edition now comprises over 600 volumes (December 1989) with nearly 190 000 pages. About 320 000 elements, compounds, and systems were described up to 1987, and about 14 000 more are added each year. Each substance is described by information about occurrence or methods of preparation, physical and structural properties, and chemical behavior. Gmelin includes many useful numerical data, graphs, and diagrams. There is extensive coverage of applied aspects and commercial manufacturing practice. Since the mid-1970s more attention has been paid to toxicological and environmental issues, uses, and applications.

The Main Volume series started in 1924 with a reporting period beginning around the middle of the 18th century and ending at best some months before publication date of the volume under consideration. This series is almost complete. Supplement volumes, which began to appear in 1937, continue the subject. The reverse of the title page for each volume shows the latest date through which literature for that volume is evaluated. A New Supplement Series started to appear in 1970. The earlier volumes are in German, but volumes produced since 1982 are entirely in English.

The Gmelin classification is based upon 71 system numbers for the various elements or combinations of them (Table 5). A compound consisting of two or more elements is indexed in the volume(s) pertaining to the element of highest system number. Formerly, librarians arranged volumes by increasing system number. The Gmelin Institute now recommends libraries to shelve the volumes alphabetically by atomic symbol rather than by system number.

The classical way of looking a compound up is to find the system number, find the appropriate volume, and then use the Table of Contents. All volumes have contents tables in English and German. Some volumes or system numbers even have particular formula indexes. All compounds up to 1987 are included in the English-language Gmelin Formula Index. The first column in this index is the empirical molecular formula in alphabetical order of the elements; C and H are not treated separately here. The second column gives the usual empirical formula as employed in the body of the Handbook text. The third column contains the system number and the fourth column the volume and page numbers. Both the Gmelin Formula Index and the Complete Catalog, which also exists in printed form, are available online via STN International.

In 1991 the Gmelin Institute will offer a Factual Data Bank as a further online product. This data bank will also include the most up-to-date facts which are not yet published in the printed Handbook.

Table 5. System numbers used in the Gmelin classification

System No.	Symbol	Element
1		noble gases
2	H	hydrogen
3	O	oxygen
4	N	nitrogen
5	F	fluorine
6	Cl	chlorine
7	Br	bromine
8	I	iodine
	At	astatine
9	S	sulfur
10	Se	selenium
11	Te	tellurium
12	Po	polonium
13	B	boron
14	C	carbon
15	Si	silicon
16	P	phosphorus
17	As	arsenic
18	Sb	antimony
19	Bi	bismuth
20	Li	lithium
21	Na	sodium
22	K	potassium
23	NH_4	ammonium
24	Rb	rubidium
25	Cs	cesium
	Fr	francium
26	Be	beryllium
27	Mg	magnesium
28	Ca	calcium
29	Sr	strontium
30	Ba	barium
31	Ra	radium
32	Zn	zinc
33	Cd	cadmium
34	Hg	mercury
35	Al	aluminum
36	Ga	gallium
37	In	indium
38	Tl	thallium
39		rare earths
40	Ac	actinium
41	Ti	titanium
42	Zr	zirconium
43	Hf	hafnium
44	Th	thorium
45	Ge	germanium
46	Sn	tin
47	Pb	lead
48	V	vanadium
49	Nb	niobium
50	Ta	tantalum
51	Pa	protactinium
52	Cr	chromium

System No.	Symbol	Element
53	Mo	molybdenum
54	W	tungsten
55	U	uranium
56	Mn	manganese
57	Ni	nickel
58	Co	cobalt
59	Fe	iron
60	Cu	copper
61	Ag	silver
62	Au	gold
63	Ru	ruthenium
64	Rh	rhodium
65	Pd	palladium
66	Os	osmium
67	Ir	iridium
68	Pt	platinum
69	Tc	technetium
70	Re	rhenium
71		transuranium elements

HCl

$ZnCl_2$

$CrCl_2$

$ZnCrO_4$

3.3.3   The Dictionary of Organic Compounds

The Dictionary of Organic Compounds is a successor to the dictionary first compiled by HEILBRON and BUNBURY in 1934. The fifth edition was published in five volumes in 1982 [3.22]. Annual supplements have been published since then. The work is indexed by name, molecular formula, heteroatom, and CAS Registry Number. The main work has 50 000 entries covering 150 000 compounds. Each annual supplement has a further 3000–4000 entries.

The dictionary is selective in coverage: it describes the 5% of compounds that are most widely known and used. Selection is made by industrial and academic experts.

Structural formulae, names, CAS Registry Numbers, physical, chemical, and (where available) biological properties, and literature references are listed for each compound.

The dictionary is phototypeset from a sophisticated electronic data base [3.23]. The chemical structure diagrams are currently artworked and scanned photoelectronically but work is in progress on a new methodology for linking structures and text [3.24].

The Dictionary of Organic Compounds, plus six other dictionaries published by Chapman and Hall, form the HEILBRON online data base on DIALOG. This data base contains 250 000 organic substances, with CAS Registry Numbers; structures can be displayed graphically using the DIALOGLINK software (see Section 8.2.9).

3.4   Sources Concerned with Chemical Reactions

Houben–Weyl [3.25]. The aims of Houben–Weyl are to deal critically and comprehensively with experimental methods; to give examples with full experimental detail; to include relevant theoretical background; and to involve experts to ensure quality. Over 80 volumes have been published so far in German.

Theilheimer. WILLIAM THEILHEIMER started to produce his renowned yearbooks, Synthetic Methods in Organic Chemistry [3.26], in 1946. Since 1974 (vol. 30) the series has been derived from the Journal of Synthetic Methods (Derwent Publications). The combination of Theilheimer and the journal forms the basis of Derwent's online Chemical Reactions Documentation Service (CRDS) [3.27], [3.28] and of the Theilheimer data base for use in-house with the reaction indexing software REACCS (see Section 10.8).

Other Reaction Compendia. A number of multivolume works are devoted to chemical reaction information [3.29]–[3.33].

3.5   Other Reference Books

Several multivolume treatises of lesser scope than Beilstein and Gmelin have been published [3.34]–[3.40].

The Merck Index [3.41], now in its 11th edition, is of particular use to the pharmaceutical and agrochemical industries. This one-volume reference book contains short descriptions of over 10 000 biologically active compounds. Computer-assisted production methods have been used for the ninth edition onwards and the work is also available online.

The CRC Handbook of Data on Organic Compounds, HODOC II contains chemical, physical, and spectral data for 30 000 compounds [3.42]. Annual supplements are printed. The publishers are considering methods of mounting the data online.

Potentially hazardous chemicals are covered in SAX's handbook [3.43]. BRETHERICK's publications on chemical hazards are equally important [3.44], [3.45]. The most complete list of toxic effects of chemicals is the Registry of Toxic Effects of Chemical Substances (RTECS) published by the U.S. National Institute for Occupational Safety and Health (NIOSH) and also available online.

The Aldrich Chemical Company, Sadtler Research Laboratories, and the Royal Society of Chemistry all produce hard-copy libraries of spectra. Analytical data are covered in more detail in Chapter 9.

Fuller lists of reference works concerning physical properties, analytical chemistry, and safety and related data are given in [3.1]–[3.8].

4   Gray Literature

4.1   Introduction

Gray literature, otherwise termed "nonconventional", "fugitive", "informal", "ephemeral", "invisible," or "underground", is literature which is not available through normal bookselling channels and which has characteristics such as small circulation and poor bibliographic control. It includes reports, theses, conference papers, preprints, official publications, translations, house journals, trade literature, and working papers. However, not all the literature in these categories is "gray". For example, conference proceedings may be published as books, many official publications are commercially available, and many journals are translated cover to cover.

Translations have been discussed in Section 1.6 and supplementary material in Section 1.8.

4.2   Characteristics [4.1]–[4.3]

Nonconventional documents often have small print runs because they are intended for a small audience. They may be produced very rapidly, for example by in-house publication, but often have variable standards of editing, legibility, and physical presentation. The issuing organization is likely to be a university, research institution, government department, public body, or commercial agency. The documents are usually poorly publicized, have inadequate bibliographic control (e.g., author, title, producer), and bear unusual codes or identification numbers. Gray literature is not easily available to the average library. It is more likely to be produced in a local language than the conventional literature. Sometimes it comes in a format (e.g., microfilm or microfiche) unacceptable to the average user. A nonconventional document is frequently large and may be an issue of a serial appearing at irregular intervals. The cost to libraries of tracing, reproducing, and distributing such documents is very high.

Exploitation of the huge body of knowledge in the much underused gray literature would make sense. Many problems could be overcome if the producers realized the value of the information and improved its accessibility [4.1], [4.4], [4.5]:

1) Documents should be produced to better physical and bibliographic standards.
2) Producers should be less restrictive about what is released, should indicate the confidentiality of documents, and should widely publicize only that material which is for unlimited distribution.

3) Copies should be sent to appropriate secondary services and national gray literature centers; and to national depositories, copyright libraries, and specialist collections which can provide bibliographic and physical access to the documents concerned.
4) Producers should have larger print runs to meet the demand generated by more publicity.

4.3   Organizations Specializing in Gray Literature

In view of the problems described in Section 4.2, the tasks of collection, bibliographic control, and circulation of gray literature fall upon specialized organizations.

United States. One the best known centers is the National Technical Information Service (NTIS) in the United States. This government agency has been acquiring, recording, reformatting, and promoting American research reports since the late 1960s. Over 2×10^6 reports are available from NTIS and about 70 000 new ones are added annually. They contain the results of U.S. and foreign government research and developments that have not generally been previously published elsewhere. They are available in hard copy and on microfiche and are sold worldwide directly or through appointed national agents. NTIS lists the documents fortnightly in the Government Reports Announcements and Index.

Other U.S. government agencies produce printed and online secondary services for thousands of reports every year: the Department of Energy (DOE), the National Aeronautics and Space Administration (NASA), and the Educational Resources Information Center (ERIC).

Europe. The Federal Republic of Germany has a decentralized system for the provision of literature, including gray literature [4.6]. Specialist libraries such as the Technische Informationsbibliothek (TIB) in Hannover and the Central Agricultural Library in Bonn have national responsibility for providing services in their subject areas.

The Centre de Documentation of the Centre National de la Recherche Scientifique (CNRS) in France and the British Library Document Supply Centre (BLDSC) have an interest in gray, as well as in conventional literature. The British Library has been collecting gray literature since the late 1960s and has a vast collection of reports, theses, translations, supplementary publications, local government documents, and semipublished conference proceedings [4.7], [4.8]. It publicizes its holdings in British Reports Translations and Theses, Index to Conference Proceedings Received, and Current Serials Received.

One of the most significant developments is the establishment of the System for Information on Gray Literature in Europe (SIGLE) [4.1], [4.5], [4.9]–[4.12]. Details of the literature are contained in an online data base (on BLAISE-LINE, Sunist, and STN International) produced by national centers in Belgium, France, Italy, Luxembourg, The Netherlands, the Federal Republic of Germany, and the United Kingdom. The centers belong to an association known as the European Association for Gray Literature Exploitation (EAGLE). SIGLE holds information on ca. 150 000 documents covering pure and applied sciences, and technology since 1980, and economics, social sciences, and humanities since 1984. The data base increases by 30 000 records per year.

International Agencies. The International Nuclear Information System (INIS) of the International Atomic Energy Agency (IAEA) microcopies the full text of non-conventional documents and makes them available on demand. It supports this service with the bibliographic tool INIS Atomindex.

The Food and Agriculture Organization (FAO) has a document supply service for reports and produces a hard-copy index, AGRINDEX, through the International Information System for Agricultural Sciences and Technology (AGRIS) [4.13].

Many publications of the World Health Organization (WHO) can be obtained through booksellers and standard channels but its production of internal, unpublished documents is disseminated through the bimonthly bulletin WHODOC: Index to WHO Technical Documents.

The International Agency for Research on Cancer (IARC) handles both conventional and gray literature.

The International Federation of Library Associations (IFLA) Office of International Lending and the Worldwide Network of Agricultural Libraries (AGLINET) supply documents through interlibrary loan schemes.

Commercial Organizations. Congressional Information Service in the United States indexes, abstracts, and microfilms all U.S. Congress publications. Micromedia index Canadian federal, provincial, and local government publications, as well as reports of research institutes and professional associations. Chadwyck Healey in the United Kingdom indexes non-HMSO (Her Majesty's Stationery Office) publications.

4.4 **Reports** [4.5], [4.9]

Bibliographic control of report literature is better in the United States than elsewhere (NTIS, NASA, and DOE, see Section 4.3).

In the Federal Republic of Germany the Fachinformationszentrum Energie, Physik, Mathematik (FIZ) produces the monthly Forschungsberichte aus Technik und Naturwissenschaften. National centers such as CNRS and BLDSC mentioned in Section 4.3 handle reports and contribute them to SIGLE.

The European Communities Commission produces and disseminates many reports. Some of these can be obtained through one of theEuropean Documentation Centers or from a European Communities Depository Library (see also Section 4.5).

Research reports on publicly funded research projects in Japan, as well as governmental reports concerning policy on science and technology are now included in JICST-E, a data base produced by the Japan Information Center of Science and Technology, available online on STN International. Over 4000 reports, including gray literature, are covered.

4.5   Official Publications [4.3], [4.8]

Several international organizations were mentioned in Section 4.3.

In the United States, reports from the Environmental Protection Agency (EPA), DOE, the Food and Drug Administration (FDA), the Centers for Disease Control (CDC) and others, are covered by NTIS. All aspects of U.S. government information are covered in [4.14].

Large numbers of United Kingdom Government publications appear in Chadwyck Healey's Catalog of British Official Publications not Published by HMSO. Some of this gray material is available from BLDSC.

The EUR reports of the European Communities are not generally deposited in European Documentation Centers but are microfilmed by the Commission of the European Communities and are input to the SIGLE data base.

4.6   Conference Proceedings

Conference material causes particular problems for the librarian and the research worker. It is estimated that 30–50% of conference papers are never published and those proceedings that are printed can appear up to 3 or more years after the conference has been held. Documents published before, during, and after a conference vary in terminology, contents, size, and value. Conference material is published erratically through many channels and is often not adequately refereed. It does not always report original work and much of it is not noted in abstracting journals [4.15].

The BLDSC announces its acquisitions in the Index to Conference Proceedings Received (ICP), which is also available online on BLAISE. ISI's Index to Scientific and Technical Proceedings (ISTP) is available in hard copy and online. Other useful listings include the Bibliographic Guide to Conference Publications, the Union List of Conference Proceedings in Libraries of the Federal Republic of Germany including Berlin (West) [4.16], and the Samkatalog over nyanskaffat Konferenstryck.

The Conference Papers Index (CPI) produced by Cambridge Scientific Abstracts is available online on DIALOG. Each meeting is assigned a unique number which, where available, matches that used in the hard-copy World Meetings publications [4.17]–[4.19] compiled by the World Meetings Information Center.

More than 1.4×10^6 references to meeting abstracts are included in SciSearch, the online version of ISI's Science Citation Index.

"Conference Proceeding" is a searchable designation for Chemical Abstracts online (on several host computers). Analytical Abstracts offers a keyword-indexed Conference Title field for easy access to meeting publications in analytical chemistry.

4.7   Theses [4.5]

The majority of American dissertations are abstracted in Dissertations Abstracts International published by University Microfilms International (UMI). Section B and its indexes concern science and engineering. The same publishers also produce Comprehensive Dissertations Index 1861–1972, American Doctoral Dissertations and Masters Abstracts. Many universities make their dissertations available through UMI but some have to be approached directly.

In the United Kingdom, BLDSC is the main source of most post-1970 British doctoral theses. The theses are kept on microfilm. The University of London provides a service for its own theses.

In France, theses have been recorded since 1884 in the Catalogue des Thèses et Écrits Academiques and Supplement D of Bibliographie de la France. PASCAL Explore – E99-Congrès, Rapports, Thèses (formerly Bulletin Signalétique-Section 401-Congrès, Rapports, Thèses) lists French reports and other gray literature originating in France. French scientific theses are available from CNRS.

German theses are included in the Jahres- verzeichnis der Hochschulschriften der DDR, der BRD und West Berlin [4.20]. Scientific theses since 1983 (F.R. of Germany) appear in Forschungsberichte aus Technik und Naturwissenschaften.

French, British, and German theses are included in SIGLE.

Canadian theses are available through the National Library of Canada in Ottawa, which has published the listing Canadian Theses since 1984.

In the Soviet Union, dissertations are only occasionally cited but are mostly published in summary form as VINITI papers.

5   Business and Economic Information

5.1   Introduction

Business information can be loosely defined as the information that is required by a business to give it competitive advantage in the marketplace. This encompasses a broad range of information on companies, markets, products, current affairs, legislation, economics, and finance, for which there are a great many printed and electronic sources.

Directories and handbooks are the "first-stop" guides to sources, e.g., [5.1], [5.2]. Essential primary publications include newspapers, trade and business periodicals, statistical publications, company annual reports, and market research reports. External agencies such as research associations, government departments, local authorities, independent information brokers, embassies, chambers of commerce, and stock exchanges are further important suppliers of business information [5.3].

Business information must be up-to-date for competitive advantage and in this respect online data bases have made a most significant contribution. Online business information has been slower to develop than the scientific and technical data bases. However, in recent years the number of business data bases has increased enormously. Financial, company, news, and market information have become higher revenue earners than the scientific and technical data bases and have an even larger growth potential.

Development in the business data-base market has been geared towards the provision of full-text data bases with frequent updating. Especially in the financial sector, information is continuously updated and available to the user as soon as it appears. The previous domination of information produced by and for the North American markets has been challenged by the emergence of European data bases. Europeans now have access to online information that is tailored to their own requirements and markets. This trend has been accelerated by the advent of the Single European Market. Details of the growing numbers of hosts and data bases available to the online user of business information are given in [5.4], [5.5].

More recent developments in optical disc technology mean that many data bases are also commercially available on CD-ROM (see Section 6.3). Information in the form of text, graphics, images, or data may be downloaded to a microcomputer's memory and manipulated with appropriate software. Commercially available CD-ROMs of relevance to the business information user include text of directories, newspapers, and company data. At the time of writing, the market is expanding, but CD-ROM is still at a fairly experimental stage.

5.2 Company Information

Company information can range from simple needs (e.g., the name and address of a company, or its directors) to more detailed information about financial performance or international activities (acquisitions and mergers).

A company's own annual reports and accounts provide authoritative information on financial data, turnover, trading profits before and after tax, earnings, dividends, and capital expenditure. This information originates from legal requirements necessitating companies to deposit certain documents with appropriate authorities or from the need to disclose certain financial information so that shares can be traded on the stock exchange. The quality and quantity of the information can vary from country to country. The United Kingdom, Republic of Ireland, Denmark, Greece, and Luxembourg, for example, all have centralized registration systems but companies in other countries register with a local registrar or regional chamber of commerce.

Company annual reports and company information from government agencies such as the Securities and Exchange Commission (USA) or Companies House (UK) are the most important sources of primary information. Providers of company information, including Inter-Company Comparisons (ICC), Jordans, Extel, and Dun and Bradstreet, all supply information based on these data.

Directories. In addition to the above sources, large numbers of directories and yearbooks give basic company information. Amongst the more prominent are the Kompass directories which are available for most European countries. Listed below are some key publications that provide a useful starting point for most enquiries. Further sources can be found in the Directory of Directories (Gale Research Co.), Current European Directories (CBD Research), and Current British Directories (CBD Research).

International
International Stock Exchange Official Yearbook, Macmillan
Chemical Industry Directory, Benn Business Information Services
Principal International Businesses, Dun and Bradstreet
Directory of Multinationals, Macmillan

Europe
Chemical Company Profiles: Western Europe, IPC Industrial Press
Europe's 15000 Largest Companies, ELC International
Major Chemical and Petroleum Companies of Europe, Graham and Trotman
Major Companies of Europe, Graham and Trotman

Federal Republic of Germany
Handbuch der Deutschen Aktiengesellschaften, Verlag Hoppenstedt, gives detailed financial information on 2500 public companies
Handbuch der Gross-Unternehmen, Verlag Hoppenstedt, describes 22000 major companies with over 100 employees or a turnover of DM 10×10^6
West German Middle-sized Companies, Verlag Hoppenstedt

Scandinavia
Major Companies of Scandinavia, Graham and Trotman, lists 1000 Scandinavian companies

United Kingdom
CRO Directory (on microfiche)
Financial Times Industrial Companies, Volume II Chemicals, Longman Group
Kelly's Business Directory, Kelly's Directories
Key British Enterprises, Dun and Bradstreet
Macmillan Top 20 000 Unquoted Companies, Macmillan
Sell's Directory of Products and Services, Sell's Publications

United States
Chemical Company Profiles, The Americas, IPC Industrial Press
Chem Sources USA, Directories Publishing Company
Directory of Chemical Products – United States, SRI International
Major Companies of the USA, Graham and Trotman
The Million Dollar Directory, Dun and Bradstreet, also available online
Moody's Manuals, Dun and Bradstreet
Standard and Poor's Register of Corporations, Directors and Executives, Standard and Poor's
  Corporation

India
Indian Chemical Directory, Technical Press publications

Far East
Major Companies of the Arab World, Graham and Trotman, covers 20 countries including Egypt,
  Iraq, Algeria
Asia's 7500 Largest Companies, ELC International

Japan
Diamonds Japan Business Directory, Diamond Lead Co.
Japan Chemical Directory, Chemical Daily Co.
Japan Company Handbook, Oriental Economist, gives detailed information on 1st and 2nd section
  companies

Other more specialized directories give a particular type of company information. Who Owns Whom (Dun and Bradstreet) is also available as an online data base and covers the United Kingdom, North America, Continental Europe, Australia, and the Far East. Each area has two volumes: the first lists subsidiary companies and gives their parents, the second lists parent companies and their subsidiaries. The Directory of Directors (Thomas Skinner) gives the names of directors of public and private companies. The Stock Exchange Companies (Financial Times) tables the performance of the 1000 largest listed UK companies over the past five years. It also gives details of every Unlisted Securities Market company. The Times 1000 (Times Books Ltd) ranks the 1000 largest UK companies by size and gives brief financial details; it also contains 500 European and other companies. The Register of Defunct Companies (Stock Exchange Press) is also useful and there are also many specific directories, such as Duns Guide to Healthcare Companies (Duns Marketing Services).

Many other key sources for company information are available in public libraries or by subscription. Extel Company Card Services (Extel Statistical Services) cover about 3400 quoted and unquoted UK companies and also large European, Australian, and North American companies. The cards give details of company activities, chairman's statements, balance sheets, dividend records, board members, profit and loss accounts, yields, earnings, and capital history. The service also covers Unlisted Securities and Third Market companies. The McCarthy Press Cuttings Service provides weekly press cuttings on UK, European, Australian, and North American companies. ICC Business Ratio Reports provide standard ratios for intercompany comparisons. The reports on different areas of industry, including the chemical industry, analyze up to 100 leading UK companies, giving basic data and ratios from company accounts, company addresses, directors, name of holding company, and principal activities. ICC Financial Surveys cover over 160 sectors of industry in the United Kingdom and for each company give date of accounts, turnover, total assets, current liabilities, and payments to directors. Credit Reporting Services are offered by Dun and Bradstreet, ICC, and Infocheck. They provide confidential, detailed financial analyses, and credit ratings on UK companies. Stockbrokers Reports are produced on industry sectors and individual companies.

Information Available Online and on CD-ROM. Much of the information contained in printed directories is also available online. General chemical industry data bases, such as the Chemical Business Newsbase (Royal Society of Chemistry), cover business aspects of the European chemical industry. This includes company information in the form of abstracts from annual reports, press releases, and promotional material. Specialist data bases concentrate on specific areas of company information, such as trademarks, mergers, and acquisitions.

The impetus for the development of CD-ROM products has come from the United States. Compact Disclosure (Disclosure Inc.) is the CD-ROM equivalent of its online data base and is updated monthly. Lotus Development Corporation produces Compustat data bases on CD-ROM giving investment information on U.S. companies. The information can be manipulated in-house using Lotus software. Moody's CD-ROM (Dun and Bradstreet) gives financial and business information on all U.S. Stock Exchange companies. Standard and Poors' Corporations on CD-ROM gives details of 9000 U.S. public companies. European products include European Kompass on Disc and Kompass on Disc. The former covers 300 000 European companies and is available in five languages, the latter covers 160 000 UK companies.

5.3 Products and Markets

Information on competitors' products is vital to most companies for finding detailed technical data on new products, plus sales and marketing information, or simply to find out the producer of a particular product or the product range of a particular company. On a broader scale, companies need to keep up to date with general trends in their area of the market and to have access to market research to successfully target their own research, development, and production activities.

5.3.1   Product Information

Directories. To find out which companies make a particular product, the best sources of information are directories with product indexes. A comprehensive listing of trade directories is Croner's Trade Directories of the World (Gale Research Company, 1985). The Kompass series of directories (Kompass Publishers) covers most Western and some other countries; the first volume in each country set is indexed by products and services. Key British Enterprises (Dun and Bradstreet), Kelly's Business Directory (Kelly's Directories), and Sell's Directory of Products and Services (Sell's Publications) also have product indexes and are useful sources of UK company product information. The Thomas Register of American Manufacturers (Thomas Publishing Company) is a key source for the United States. Directories can also be used for finding out which products a particular company makes, although they rarely give much detail. Some trade journals contain trade directory information and a useful guide to these is Trade Directory Information in Journals (British Library).

Trade Mark Information. The major UK source for finding out which company owns a particular trade name is UK Trade Names (Kompass Publishers). Other directories such as Sell's Directory of Products and Services (Sell's Publications) and Kelly's Business Directory (Kelly's Directories) include trade name indexes. The major U.S. source is The Trade Names Directory (Gale Research). For European and international trade mark information, local company directories with a trade name index must be used. The Kompass series is a good starting point. For Japan, Diamond's Japan Business Directory (Diamond Lead Company) has a trade mark index. Gardener's Chemical Synonyms and Trade Names is useful for chemicals. The Trade Mark Registry (State House, High Holborn, London WC1) maintains a listing of registered trade marks which can be consulted by the public for a fee. However not all UK trade marks are registered, because there is no legal compulsion for a company to do so.The Patent Office, (25 Southampton Buildings, Chancery Lane, London WC2A 1AW) produces the weekly Trade Marks Journal which advertises newly registered trade marks for the United Kingdom. The U.S. Patent and Trademark Office (The Commissioner of Patents and Trade Marks, Patent and Trademark Office, Washington D.C., 20231) can supply information for the United States.

Trade Literature. For more detailed information on a product, the company's own trade literature is useful, although it may not be easily obtainable. The British Library's Science Reference and Information Service (25 Southampton Buildings, Chancery Lane, London WC2A 1AW) keeps a large collection of trade literature, covering around 12 000 companies (mainly UK) and a range of industries. Another source of trade literature may be the appropriate trade association. The chemical industry has many such associations, the most appropriate can be found in [5.6]. Trade associations for the United States, European, and Eastern countries are given in [5.2]. The country's embassy may also be able to assist.

Press Cuttings. The McCarthy Press Cutting Service includes a classified products and services sequence in its system of cards. It covers UK quoted companies and some larger European, Australian, and North American companies. The weekly service is available on subscription, but is also held in large public libraries.

Online Sources. The user should consult the Online Business Sourcebook [5.4] or Directory of Online Data bases [5.5] for a comprehensive listing of online sources. Four useful data bases for the chemical industry are Chemical Business Newsbase (Royal Society of Chemistry), Chemical Industry Notes (Chemical Abstracts Service), European Chemical News (European Chemical News), and East European Chemical Monitor (Business International S/A). They all contain data on companies, new products, and markets.

The directory Wer Liefert Was? (Wer Liefert Was? GmbH) lists $> 1 \times 10^6$ links between products and suppliers in Europe, and is available online and on CD-ROM. It operates in five languages and its 55 000 addresses can be manipulated by the user's own word-processing software. The Thomas Register of American Manufacturers is available on CD-ROM as well as in hard copy (DIALOG and Thomas Publishing Company).

5.3.2   Market Information

Guides to Market Research Reports. Market research can be carried out in-house or contracted out to a market research company. Both options are expensive and it is well worth first finding out whether the information required already exists as a published market research report. There are several guides to market research reports:

Marketing Surveys Index, MSI (UK) Ltd (updated monthly with a very detailed subject index, covering products as well as general areas; also available online)

Market Research: a Guide to British Library Holdings, British Library (all reports listed are available to the general public)

Marketsearch, British Overseas Trade Board

FINDEX: the Directory of Market Research Reports, Studies and Surveys, U.S. National Standards Association (also available as an online data base, covering 12 000, mainly U.S., reports)

The major producers of market research reports are Inter-Company Comparisons (ICC), Mintel Publications, the Economist Intelligence Unit, Frost and Sullivan, Jordan & Sons, Arthur D. Little, and Euromonitor. Many of these concentrate on consumer and retail markets. For coverage of the chemical industry, it is necessary to turn to more specialized publishers such as Chem Systems International Ltd, SRI International, Kline, the Freedonia Group, and IAL Consultants Ltd.

A typical report contains a detailed review of the industry structure, major companies involved, market size and trends, product sales by volume, value and market share, recent developments, and future prospects. Some reports may deal solely with technology and not with markets. The following list gives typical market research reports covering the chemical industry:

Biotechnology Products, Key Note Publications (1988), covers UK
Britain's Plastics Industry, Jordan & Sons (1987), covers UK
Chemfacts: United Kingdom, Chemical Intelligence Services (1987), covers UK
Chemfacts: Polypropylene, Chemical Intelligence Services (1987), worldwide coverage
Chemfacts: Polyethylene, Chemical Intelligence Services (1987), worldwide coverage
Chemfacts: West Germany, Chemical Intelligence Services (1987), covers F.R. of Germany
Custom Chemical Synthesis (1987), IAL Consultants, covers France
Dyes and Organic Pigments (1988), Freedonia Group, covers USA
Electronic Chemicals (1987), IAL Consultants, covers France
Ethylene Oxide and Derivatives (1988), Freedonia Group, covers USA
Fertiliser Outlook (1988), Freedonia Group, covers USA
UK Chemical Manufacturers and Distributors (1987), ICC Financial Surveys, covers UK

Market research periodicals include Marketing in Europe (monthly, Economist Publications), Market Research Great Britain (monthly, Euromonitor), and Marketing Trends (semi annual, AC Nielson). The last is available in British, American, German, Spanish, and Portuguese editions. These contain topical marketing information and are a useful means of keeping abreast of current trends.

Online Sources. ICC International Business Research (ICC Information Group) gives abstracts of all ICC's Keynote Market Reports. Arthur D. Little Online (Arthur D. Little) gives full text for around 80% of their publications, including industry forecasts and market reviews. The data base's coverage is mainly United States, although some is international. The Media Expenditure Analysis Ltd. (MEAL) data base gives information on advertising and expenditure in the United Kingdom.

Marketing Associations. If published sources fail to provide the necessary information then the user can commission a market research company to produce a report. Examples of marketing associations able to give advice and further information are:

Association of Market Survey Organisations Ltd
  60 Kenilworth Road Ltd
  Leamington Spa
  Warwickshire CV32 6JY
Industrial Marketing Research Association
  11 Bird Street
  Lichfield
  Staffs WS13 6PW
American Marketing Association
  22 South Riverside Plaza
  Suite 606
  Chicago IL 60606
Japan Marketing Research Association
  No. 20 Sankyo Building
  11−5 Iidabashi-3-chome
  Chiyoda-ku
  Tokyo 101
TMO Consultants
  22 rue de Quatre Septembre
  75002 Paris

5.4   News Services

Scanning newspapers and business periodicals has been the traditional way of keeping up to date with business news and developments. Periodicals such as Chemical Insight (published semimonthly by Hyde Chemical Publications) cover the chemical industry on an international basis, whilst others like Japan Chemical Week help to monitor events in specific countries. Timely access to new information on competitor activity and other market developments clearly help a business to make informed decisions, but delays in receiving foreign publications can result in news being stale by the time it is read.

The 1980s have seen the development of full-text online data bases designed to provide rapid access to industry news. Reuter Textline (previously owned by Finsbury Data as Textline) contains abstracts and full text articles from over 1400 newspapers and journals. It covers the major UK dailies, various provincial newspapers, European and international press, and trade press. It also includes all Reuter wires from journalists and has a specialist section for chemicals. It has records from 1980 onwards and is updated daily. NEXIS (Mead Data Central International) is a full-text data base that is updated rapidly and has four main files: NEXIS Magazines, NEXIS Newspapers, NEXIS Newsletters, and NEXIS Wire Services. It is biased towards U.S. sources, but is expanding its European coverage. PTS Promt (Predicasts Inc.) covers company and government news, market information, and details of new products and processes. Coverage is international with a U.S. bias.

Profile Information (Profile Information, a subsidiary of Financial Times Information Online Ltd) has vast files of full-text sources covering a wide range of company, business, marketing, and industry news. Sources include The Washington Post, Financial Times, Asahi News Service, BBC Summary of News Broadcasts, and TASS Newswire. Infomat (previously BIS-Infomat and now also produced by Predicasts) covers around 500 European and international newspapers and journals. It is not full text but produces informative abstracts and aims to give good coverage of the European Single Market.

Nikkei Telecom Japan News and Retrieval is a real-time data base which displays news as soon as it is filed. It gives English-language coverage of Japanese newspaper items and articles, often several hours before the newspaper is published in Japan. There are also files covering Tokyo Stock Market and Money Market figures. A broadcasting mode is available, whereby real-time news is displayed as soon as it comes into Nikkei Telecom.

Chemical Business Newsbase (Royal Society of Chemistry), covers trends and current affairs in the European chemical industry and end markets. Chemical Industry Notes (Chemical Abstracts Service) covers worldwide business news on the chemical processing industries, giving information on production, pricing, sales, products, processes, and corporate activities.

Routine monitoring of an appropriate data base allows the user to keep abreast of all developments in a certain business area in a fraction of the time it would take to scan manually all the appropriate publications. The full range of online news data bases can be found in [5.3]

CD-ROM is less useful for news information than online services due to delays in updating but this is an expanding area. The National Newspaper Index (Information Access Company) indexes five major U.S. newspapers. It covers four years of data and is updated monthly. Newspaper Abstracts Ondisc (University Microfilms International) gives brief abstracts for a range of U.S. newspapers. ABI-INFORM, (University Microfilms International) contains abstracts of articles from over 800 business journals. Its sister product, Business Periodicals Ondisc, gives access to the full text of the current and backfile issues of 300 business and management journals.

Some of the newspapers and periodicals of interest to the chemical industry follow. For more detailed listings for individual countries, see [5.2].

United Kingdom (general business)
Financial Times
The Guardian
The Independent
The Times
The Observer
The Accountant (monthly, Lafferty Publications Ltd. London)
Business (monthly, Business People Publications, London)
Campaign (weekly)
The Economist (weekly, Economist Newspapers Ltd, London)

United Kingdom (chemical industry)
British Journal of Pharmaceutical Practice
British Polymer Journal
Chemical Engineer
Chemistry & Industry
Chemistry in Britain
European Chemical News
Fertiliser International
Laboratory Products
Industrial Chemistry Bulletin
Pharmaceutical Business News
Practical Biotechnology
Process Engineering
Process Equipment News
Scrip World Pharmaceutical News
Speciality Chemicals

United States (general business)
Journal of Commerce
New York City Tribune
New York Post
New York Times
The Wall Street Journal
Barrons (weekly, Dow Jones)
Business Week (weekly, McGraw Hill)
Forbes (fortnightly, Forbes Inc.)
Fortune (monthly, Time Inc.)
Harvard Business Review (bimonthly)

United States (chemical industry)
American Chemical Society Lab Guide
Butane–Propane News
CEC–The Process Industry Catalog
Chemical Business
Chemical and Engineering News
Chemical Industry Product News
Chemical Marketing Reporter
Chemical Week
Industrial Chemical News
Soap, Cosmetics, Chemical Specialties

European and International coverage (chemical industry)
Chemie Anlagen & Verfahren Europe
Chemical & Engineering News
Chemical Week
Chemical Products
Chemische Industrie International
European Chemical News
European Plastics News
International Labmatic

5.5 Legal Information

Each country has its own ever-changing legal system and there are obvious problems in finding legal information for an unfamiliar territory. Interpretation of the law is very much the concern of the experts and a major consideration is the currency of the information. Many important reference works are published in loose-leaf format so that they can be constantly updated. Online information is especially valuable but the most prominent online legal data base, LEXIS (Mead Data Central), is designed for the exclusive use of lawyers.

Company law usually comes to the fore in a take over, merger, or flotation. The law affecting day-to-day business covers such items as trade descriptions, weights and measures, sale of goods, industrial relations, health and safety, planning, and environmental law.

For most industries in the United Kingdom the primary source of information on new legislation and cases is the relevant trade association or trade publications. For more detailed study the user should consult up-to-date text books or Halsbury's Laws of England (Butterworths).

The same comments with regard to trade associations apply to EEC law. Useful sources of information on European legislation are to be found in [5.7], [5.8]. Local law in each member state often has as much impact as community law.

Each state of the United States has its own public, criminal, and private law, both common law and statutory. Unifying influences include the decisions of the Supreme

Courts and the lower federal courts and the requirement of the Constitution that every state shall give full faith and credit to the public acts, records, and judicial proceedings of every other state. States tend, however, to adopt uniform laws on specific issues such as sales and commercial transactions. It is even more important to obtain expert legal advice in the United States, if only because public authorities and other companies are likely to have lawyers readily available. The two main online data bases in the United States are LEXIS (Mead Data Central) and WEST-LAW (West Publishing).

5.6   Economics and Finance

To do business successfully in any country, information on its economic background and current economic and financial situation is vital. Every country has its own unique economic and financial situation and information sources therefore differ. A good starting point is one of the regional or international organizations that collect and disseminate information from various sources.

5.6.1   Hard-Copy Sources

The United Nations publishes a vast amount of data and a useful guide is UN-DOC Current Index: United Nations Document Index (bimonthly, United Nations, New York). Publications such as the World Trade Annual, Yearbook of National Account Statistics, Monthly Bulletin of Statistics, and The Growth of World Industry are of interest to businesses in the chemical industry.

In the United Kingdom Her Majesty's Stationery Office (HMSO) is one of the major information providers. Economic Trends, published monthly, charts statistics which give the background to current UK economic trends. Financial Statistics is a monthly compilation of key financial and monthly statistics, including government income and expenditure, public sector borrowing, banking statistics, money supply, institutional investment, company finance and liquidity, security prices, and exchange and interest rates. UK National Accounts (The "Blue Book") gives detailed estimates of production, income, and expenditure for the UK. UK Balance of Payments (The "Pink Book") covers balance of payment statistics for the preceding eleven years. Business Monitors are arranged by industry sector and give statistics on sales of UK manufactured products. They include import and export data, producer price indices, and employment figures. The Economic Commission for Europe publishes the Economic Survey of Europe and the Economic Bulletin of Europe annually. The Economic Commission for Asia and the Far East produces The Statistical Yearbook for Asia and the Far East. The Economic Commission for

Africa produces Foreign Trade Statistics of Africa and African Economic Indicators.

The EEC is a prolific producer of economic information in nine languages. Reference texts, such as Europe in Figures, cover the socio-economic situation in the EEC. The Official Journal of the EEC has several series: Series A, economic trends; Series B, business and computer survey results; Series C, communications; and Series L, legislation. Data for Short-Term Economic Analysis is published eleven times a year and is a review of the main quantitative data in relation to the community and its member states, covering economy, employment, industrial production, prices, finance, and balance of payments. National Accounts ESA–Aggregates is useful for country comparisons, since it covers accounts, surveys, and statistics of the United States, Japan, and European countries, harmonizing the results in accordance with the European System of Integrated Economic Accounts (ESA). The Consumer Price Index, Money and Finance, and ECU–EMS Information are all EEC periodicals offering current financial information.

The Organization for Economic Cooperation and Development (OECD) produces useful sources of economic information on Canada, Japan, the United States, Australia, New Zealand, Yugoslavia, and Western Europe. The OECD Economic Survey Series annually analyzes developments, prospects, demand, wages, money and capital markets, balance of payments, and government policies, with an issue for each country covered. OECD Main Economic Indicators (monthly) includes records of industrial production and consumer price indices. The Chemical Industry (an annual) gives production, price, and investment data for the major branches of the chemical industry.

Other OECD publications can be accessed via its Catalog of Publications, published every two to three years and updated in between by supplements.

The International Monetary Fund (IMF) produces the fortnightly IMF Survey which details the Fund's international banking activities; its Annual Report: Exchange Arrangements, and Exchange Restrictions provides information on foreign exchange rate systems.

The Economist Intelligence Unit (EIU) provides a wider scope of economic and political information. Its Country Reports (quarterly) monitor political, economic, and business developments in over 165 countries. Country Profiles are produced annually for each country, giving a survey of political and economic background. They are a useful starting point for the user embarking on an investigation of a company or industry in an unfamiliar country. Economic Prospects provides in-depth forecasting reports for selected countries. EIU Business Updates are monthly summaries of economic news forecasts and statistics on the major OECD countries. European Trends is a quarterly analysis and discussion of developments in the EEC.

5.6.2 Online Sources

For a full listing of data bases of economic and financial information, the user should consult [5.2], [5.4].

International. Data Resources Inc. produce DRI World Forecast, DRI External Debt, DRI International Cost Forecasting, and DRI Current Economic Indicators. The last of these sources provides major financial and economic indicators for 35 countries (including 15 developing countries). It is updated monthly and dates back to 1960. The IMF produces IMF International Financial Statistics and IMF Balance of Payments. The OECD Main Economic Indicators are available online, providing a respected source of data on economic affairs of OECD member countries. For trade statistics, Tradstat (Data-Star) is believed to cover 85% of world trade.

United Kingdom. The Bank of England Databank (Bank of England) gives access to a wide range of time series on financial indicators for the United Kingdom allowing trends to be studied and predictions to be generated. The Central Statistics Office (CSO) Data Bank has a wide coverage of monthly, quarterly, and annual statistics and includes data from the hard-copy versions of UK National Accounts and UK Balance of Payments.

Europe. Probably the most important data base is Cronos–Eurostat, produced by the European Communities Statistical Office. It covers over 900 000 time series of economic data, including general statistics, national accounts, industry and services, and foreign trade. Other major data bases include DRI Europe (Data Resources Inc.), which provides comprehensive European indicators, and DRI European Forecast.

United States. Citibase (Citicorp Data Base Services) is a U.S. macroeconomic data base and contains 5000 time series, most going back to 1947. Other data bases include DRI CFS Cost and Industry (DRI Europe Ltd) which includes export and import price indices by commodity; DRI Long-Term Industry Forecast (DRI Europe Ltd) which gives ten year forecasts for U.S. industrial activity; and MMS Equity Market Analysis (MMS International) which offers analysis of fiscal and trade policy, and economic factors affecting U.S. equity prices.

Other Countries. Eastern Bloc Economic Statistics (Vienna Institute for Comparative Economic Studies) contains over 7000 time series of economic data for Bulgaria, Czechoslovakia, the former German Democratic Republic, Hungary, Poland, Romania, the Soviet Union, and Yugoslavia. It is available through Reuters. DRI Asian Forecast (DRI Europe Ltd) covers eleven countries and gives detailed projections of economic and financial conditions. Of similar use are DRI Latin American Forecast and DRI Middle East and African Forecast.

6 Information Technology—Hardware

In the twentieth century there has been an exponential growth in the published literature. Chemical Abstracts recorded about 12 000 abstracts in 1907 but nearly 490 000 in 1989. Chemical companies have also experienced an enormous increase in the number of internal records; very large collections of chemical structures and related property data have been established in-house. It would not have been possible to manage these huge quantities of data without computers or the great advances made in information technology since the 1970s.

This brief introduction to information technology covers only those aspects of greatest relevance to online searching, and to in-house documentation and chemical structure handling systems [6.1], [6.2].

6.1 Digital Computers

A computer's memory consists of a large number of electronic switches represented as 1 (on) or 0 (off). A combination of switches, or bits, can be used to represent numbers, letters, and other data to be handled by the computer. The American Standard Code for Information Interchange (ASCII) was created to assign binary codes to data. Eight bits, forming a byte, are used to represent each character. (Large IBM machines use the extended binary coded decimal information code, EBCDIC.)

Modern computers use semiconductor memory, silicon usually being the basic material. The number of integrated circuits that can be stored in a single chip has increased rapidly from tens of transistors in 1973 (small scale integration, SSI) up to millions of transistors (ultralarge scale integration, ULSI) in 1990.

A computer consists of a central processing unit (CPU), which is an assembly of electronic circuits linked to data storage devices, input devices, and output devices.

First-generation computers (1946–1955) were based upon valve technology. They were very large and had very poor performance. Programming was laborious using machine code (binary) or an assembly level language (a symbolic version of machine language). Second-generation computers (1956–1963) used transistors and had compilers. Programs could be written more quickly and conveniently in a high-level language which the compiler converted into machine instructions. Third-generation computers (1964–1981) had integrated circuits, semiconductor memories, and timesharing operating systems (see Chapter 7) and made large-scale use of disk storage. The current fourth generation of computing uses very large-scale integration, VLSI (hundreds of thousands of transistors on a single chip).

Research continues into fifth-generation computing. The Japanese Fifth Generation Project [6.3] began in 1982. It was closely followed by the British 5-year "Alvey" program (1983) [6.4]; the European Strategic Program for Research and Development in Information Technologies (ESPRIT, 1984); and initiatives sponsored by American, French, German, and East European goverments and industry [6.5], [6.6].

Advances in microelectronics could increase processing power by a factor of 1000 [6.7], [6.8]. However, physical constraints (e.g., the speed at which electrons can move and the need to dissipate heat) will eventually prevent further large increases in the performance of any one processor. Research is, therefore, also taking place into "parallel processing", the design of systems which can perform large numbers of simultaneous operations [6.9]. Some applications of parallel processing in information science are considered in Sections 8.4.4 and 10.4.4.

Another development is reduced instruction set computer (RISC) technology which makes more efficient use of the CPU by reducing the number of different machine instructions.

Studies are also being carried out into man-machine interfaces [6.10], which could provide more sophisticated methods of interaction between computers and users (Section 11.5) and into expert systems which allow inferences to be drawn from knowledge encoded in machine-readable form (Section 11.6).

The most powerful computers are termed supercomputers. Corporate large computers with hundreds of users are referred to as mainframes. Minicomputers are rather less powerful multiuser machines. Personal computers are often referred to as microcomputers (see Section 6.4).

6.2   Computer Peripherals

Peripherals are the means of getting information into and out of computers.

Printers [6.1], [6.11], [6.12]. Paper-based input (cards and tapes) is now of historical interest only but printed output continues to be popular.

Serial printers (e.g., dot matrix, daisy wheel, thermal, ink-jet) print a single character at a time. Dot matrix printers are noisy and do not produce particularly high-quality prints but they print quickly and cheaply. Daisy wheel printers are slower and more expensive than dot matrix printers, and are also noisy, but they produce prints of much higher quality. Thermal printers are much quieter but are slow and the image degrades with time and on exposure to light. Ink-jet printers can produce high-quality print at high speed but they are more expensive and less reliable than daisy wheel printers. They can print graphics and characters and are capable of high-quality color printing.

Line printers were the traditional method of producing large quantities of print from mainframe computers, printing an entire line at a time.

Laser printers can produce a whole page of high-quality print at a time; they have become commonplace despite their (comparatively) high cost.

Computer-output microfilm (COM) is considered in Section 8.4.1.

Graph Plotters [6.11]. Both drum plotters and flatbed plotters use pens on paper. They retain a market niche for special applications, especially where high quality or extensive use of color is required.

Computer Terminals. The use of teletypewriters is now rare and visual display units (VDUs) are commonplace. The VDU has a display screen, a data store, a character generator, a keyboard for data input, and a communications interface. "Intelligent" terminals have local processing capabilities.

Graphics Terminals [6.13]. In the early 1980s graphics terminals were expensive and were restricted to uses such as molecular modeling (see Section 10.7.3) and computer-aided design and manufacture (in an engineering environment). With the advent of microcomputers (see Section 6.4), most scientists now have access to graphics systems for handling chemical structures for producing graphs and charts, and for desktop publishing. An important factor is the resolution of the graphics display system. Low resolution is adequate for most home computing whereas computer-aided design and manufacture require very high resolution. Most chemical structure handling systems require intermediate resolution.

There are two main ways of producing graphics on a cathode ray tube (CRT) display: raster displays and vector displays. Raster displays work much like a television: the screen is scanned by a large predetermined number of horizontal lines that provide uniform coverage of the display space. Vector displays work by drawing lines on the screen. The latter are more expensive and are gradually being overtaken by raster technology, except for specialized uses (e.g., rotating molecules in three dimensions) where both interactivity and high resolution are required. (Interactivity refers to the speedy, effective communication between humans and machine when response is fast.)

Light pens are used to input two-dimensional visual data to a graphics terminal [6.1], [6.13]. One end of the pen houses a light-sensitive transistor which senses the illuminated pixels (picture elements) on the screen. The action of pointing the pen at the screen requires the user to hold an arm up for extended periods of time. *Graphics tablets* do not have this disadvantage since they act as horizontal "drawing pads" [6.1], [6.13]. A tablet is used in conjunction with a "pen" which has a microswitch in the tip. When the tip is depressed, the coordinates of the pen-down position are transmitted to the graphics screen.

Two-dimensional data are more commonly input to a microcomputer by means of a *mouse*, which is a small hand-held body with a freely turning ball underneath it [6.1], [6.13]. As the mouse is moved around, the movements of the ball are transmitted to the cursor on the VDU. Mechanical mice run on any flat surface; optical mice require a special pad with grid lines. Mice and related devices are useful not only for

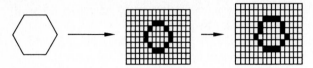

Figure 2. Creation and scale-up of a bit-map image

inputting drawings, but also for moving a cursor in selection of menu items in a menu-driven program.

Touch screens have a grid of IR beams in front of them, which can be broken by a finger or a suitable implement. They are not very precise but are useful for inexperienced users operating menu-driven software in, for example, public information systems.

Image Scanning [6.14]–[6.16]. Digital raster scanning and optical character recognition (see below) are alternatives to key punching for the conversion of a paper document into electronic form.

Image scanning uses digital facsimile transmission technology to capture a page image and convert it into a digital code that can be displayed as a visual image. However, text and numbers within an image, once stored, cannot be interpreted or processed. If the input pages are manually indexed they can be retrieved as visual images but items within each page (image) cannot be retrieved.

The images are represented electrically as a stream of very small picture elements (pels or pixels). Each pixel may exist in one of two states, black or white, represented by one bit of data. (Color and half-tone images require a more complex methodology.) The collection of individually processable pixels is referred to as a bit map (Fig. 2). Bit-mapped images occupy a considerable amount of storage space compared with text and they are usually compressed algorithmically to reduce storage requirements and increase transmission speed.

Images may also be represented by vectors. For vectorization the line structure of the image is analyzed and described in terms of its coordinates; it is specified as a set of descriptions of its component lines and shapes.

The PATDPA system which involves vectorization for further compression and efficiency is described in Section 8.2.9. Other online systems that can display images are described in the same section (see also Chap. 20).

Optical Character Recognition (OCR) [6.15]. Optical character recognition equipment (Section 11.5.2), such as that of Kurzweil, reads individual printed or typed characters and converts them into digital character codes. The text can then be indexed, edited, reformatted, or used in an information retrieval system. Optical character recogniton is much more dependent on the quality of the source document than digital raster scanning. Manual checking of input material and correction of misinterpreted characters may be necessary.

Bar Codes. Bar codes represent information as black and white bars in a logical pattern. A beam of light is shone across the bars and the reflected light is converted into voltage differences that the computer can recognize.

Bar codes have been used in library applications (e.g., to register borrowed books) and in inventory systems for laboratory chemicals to record bottle movements.

6.3   Data Storage

Textual, numeric, graphic, audio, and video data may be recorded. Comparisons of some storage media, including speeds, costs, and capacities are given in Tables 6 and 7.

Paper. The widespread implementation of office technology has not apparently decreased the the volume of paper in offices and document collections. Paper occupies a lot of space, may deteriorate in storage, and has poor security. However, paper records are necessary for some legal purposes.

Microforms [6.2]. Microform images are recorded on photographic film in roll-film format or flat, microfiche format. They may be used in computer-based information systems (see Section 8.4.1) or may be handled manually using reader – printers. They occupy about 2% of the space occupied by the original documents and therefore offer considerable savings in storage space and cost for text and graphics in records collections, archives, and libraries. However, retrieval is slow and users are resistant to the manual operations involved. Microforms are therefore best used for high-volume archiving where little retrieval is needed.

Magnetic Media. Magnetic media (tape and disks) have the advantage of being reusable: the data can be overwritten many times if required. *Magnetic tape* is

Table 6. Cost, capacity, durability, and speed of data storage media*

Medium	Capacity of one unit, Mbytes	Cost (1989), $/MByte	Removable	Durability	Speed
CD-ROM	550	0.005	yes	stable	medium
Online	unlimited	≤ 200	no		very slow
Paper	0.002	7	yes	stable	extremely slow
Tape	60	0.5–1	yes	volatile	very slow
Floppy disk	0.36–1.4	1–2	yes	volatile	medium
Hard disk	10–600	10–20	no	volatile	very fast
Bernoulli box	20–40	5	yes	volatile	fast
WORM	100–400	0.5–1.5	yes	stable	medium fast

* Reproduced with permission from a presentation given by STUART MARSON to the American Chemical Society Division of Chemical Information, Dallas, Texas, April 1989.

Table 7. Comparison of different storage systems listed according to capacity*

Storage medium	Capacity per unit, Mbits	Number of A4 pages	Density, bit/mm^2	Write/read transfer rate
One A4 page (2000 characters)	0.016	1	0.45	150 bit/s
Semiconductor memory	0.256	16	10×10^3	5 Mbit/s
Magnetic bubble memory	1	62.5	15×10^3	50 kbit/s
Magnetic disk	560	35 000	15×10^3	15 Mbit/s
Magnetic tape (dp)	720	45 000	1×10^3	10 Mbit/s
Music cassette, 60 min (analogue)	(860)	62 500	2×10^3	15 kHz
Audio disk (analogue)	(1200)	75 000	10×10^3	20 kHz
Holographic memory	10 000	630 000	1×10^6	100 Mbit/s
Compact disc	15 000	940 000	270×10^3	4.5 Mbit/s
Optical disc, 30 cm	20 000	1.3×10^6	2×10^6	10 Mbit/s
Digital optical disc, 30 cm	30 000	1.9×10^6	470×10^3	16 Mbit/s
Video tape (analogue)	(150 000)	9.4×10^6	120×10^3	8 MHz
Video play v.l.p.	(150 000)	9.4×10^6	2.7×10^6	10 MHz
Human brain	1×10^6 (long-term storage)	62.5×10^6	$(10^9/cm^3)$	1 bit/s (long-term memory) 50 bit/s (short-term memory)

* Reproduced with permission using data supplied by Polygram and Siemens [6.17].

portable, making transfer of data from one system to another possible, but the data has to be read serially, which is slow.

A *magnetic disk* has a film of magnetic material on a disk. The whole disk is rotated while the read–write heads move over it. Both fixed and removable disks are available. The disks allow random access, which is faster, and can store more data than a tape (0.5–1.2 Gbytes compared with 20 Mbytes for a tape).

Digital Optical Discs [6.18]. Optical media offer huge storage capacity at low cost. They are not prone to magnetic destruction and should not in theory deteriorate with repeated use since the disc is not physically touched as it spins in the reading device.

Technology depends on the use of a high-powered laser to create a physical change in the surface of the disc. The laser can burn a pit in the surface, create a bubble, or change the state of the surface in a more subtle way (e.g., a change from the crystalline to the amorphous state). The pattern thus created is equivalent to a series of zeroes and ones. A low-powered laser system can read the pattern by measuring light reflectance from the surface.

Three types of optical memory are of significance in information processing: compact disc-read only memory (CD-ROM), write-once read many times (WORM) discs and erasable/rewriteable discs.

CD-ROM [6.1], [6.18]–[6.26]. Compact discs were originally designed to reproduce high-fidelity sound for use in the home. The market success of CD audio, and the similarity of production technology for CD-ROM, gave CD-ROM a head start in information processing.

The manufacturing process involves premastering, mastering, and replication. In premastering the data are converted to a master tape. The tape is read by a minicomputer and a glass master disc is cut. The disc is coated with a photoresist and a laser beam creates a pattern corresponding to the data to be encoded. An etching process then produces a series of pits and "lands" (holes or no holes). The glass master is converted to a "stamper" in a complex duplication process. The CD-ROMs are produced from the stamper by injection molding of a polycarbonate resin. Each CD-ROM is finally given reflective and protective coatings.

About 600 Mbytes of data (equivalent to 270 000 typed A4 pages) can be stored on one 12-cm (4.72 inch) CD-ROM. In 1986 it was estimated that it cost about $3000 to produce a master but only about $15 for each CD-ROM. Costs have been decreasing ever since. However, the complexity of the production process means that CD-ROM is not suitable for data that need regular updating.

Since CD-ROM is a read-only medium, it cannot be repeatedly overwritten as can a magnetic tape or disk. This, however, is an advantage for the integrity of storage of data bases and documents. CD-ROMs have slower access times than magnetic disks.

A CD-ROM can be used by only one user at any one time and is read by a "player" attached to a microcomputer. Some applications of CD-ROM are considered in Sections 8.2.10 and 8.3.2.

Variations on CD-ROM such as compact disc interactive (CD-I) [6.22] for storing text, still pictures, and audio or digital video interactive (DVI) [6.23], which can handle motion pictures, text, graphics, and audio, are not yet widely applicable in information and documentation.

WORM. This differs from CD-ROM in that the disc is supplied as a blank and the user writes in his own information. Once the disc is written, the data cannot be modified or added to. WORMs are used by reports collections and archives which have to store vast numbers of documents (see Section 8.4.2). The discs are commonly 12 inches in diameter and about 100 of them can be used together in digital "juke boxes", giving huge storage capacity.

Erasable discs are only just emerging from the laboratory and are still very expensive. They are more applicable to computer science than to electronic publishing or records management [6.27].

Other Optical Storage Media. In 1988 ICI introduced a new, inexpensive optical data storage medium called Digital Paper [6.28], [6.29] consisting of an extremely thin, flexible polymer film fabric coated with an IR-sensitive dye. Information is stored in much the same way as optical discs but at much higher densities and the material can be made into sheets like paper, stamped out as discs, or wound onto cassettes as optical tape. A 12-inch reel of 35-mm Digital Paper could store 1000 Gbytes of data (as much as 1600 CD-ROMs or 1×10^9 sheets of paper).

Drexler Technology's laser cards carry data on a device similar to a credit card [6.30], [6.31].

Semiconductor and Bubble Memories. Semiconductor-based storage devices include *random access memory* (RAM) and *read only memory* (ROM) integrated circuits.

ROM contains data which are to be permanently stored and need to be directly addressed and rapidly accessed by the CPU. The data are not lost when the machine is switched off. RAM differs from ROM in that all the data in RAM are lost once the machine is switched off.

Bubble memory is the subject of much research [6.1], [6.2]. The name derives from "bubbles" (minute cylindrical magnetic domains) that are propelled through a thin film of magnetic material. The presence of a bubble signals a one and the absence of a bubble represents a zero. Permanent storage is possible within the CPU rather than on an external device; the stored data are not lost when the power is switched off (unlike data on RAM). Bubble memories are compact, have no electromechanical parts, and (unlike disks) use very little power. Access is faster than with magnetic storage media but capacity is lower.

6.4   Microcomputers [6.32], [6.33]

First-generation microcomputers were relatively slow and had low-resolution graphics; many had no graphics capability. In general they were used for local data handling and were not linked as terminals to mainframes or in local area networks. With the evolution of terminal emulation software and communication boards that facilitated these links, microcomputers replaced dumb terminals as the preferred method of connecting to mainframe computers. Soon after, networking software and hardware allowed microcomputers to be grouped into local area networks. However, for computationally intensive tasks requiring high-resolution graphics (e.g., molecular modeling and computer-aided design), the personal computer had to be enhanced by special graphics circuit boards, screens, and additional coprocessors (i.e., extra boards). The scientific workstation was designed to fill this gap.

Scientific workstations are characterized by high-resolution graphics, either raster (e.g., Silicon Graphics), or vector (e.g., Apollo Domain). They offer a choice of up to 256 colors, window-oriented user environments (see Section 7.5), and multitasking operating systems. They often have high-bandwidth network links to other workstations or mainframes and shared computational facilities.

Third-generation microcomputers based on 32-bit microprocessors have the computational power of earlier minicomputers. (The terms 8-bit, 16-bit, 32-bit etc. refer to the size of normal machine instruction that the CPU can process). Examples are the IBM PS/2 and the Apple Macintosh II, equipped with Intel 80386 or Motorola 68020–68030 chips, respectively. These machines can process 4×10^6 instructions per second, comparable to the processing power of a DEC VAX 11/780 minicomputer of the mid-1980s. They offer high-resolution graphics at a price an order of magnitude lower than that of the typical scientific workstation of the early 1980s.

Scientific workstation manufacturers responded with newer models at lower prices to compete with top-range microcomputers. Mainframe and minicomputer manufacturers also started to enter the workstation market. The distinction between the top end of the personal computer range and the bottom end of the workstation range is thus blurred.

7 Information Technology— Software and Environments

7.1 Systems Software and Application Software

Software is a generic term covering the concepts, procedures, and instructions which cause computer systems to perform useful tasks. It is usually thought of in terms of individual programs and integrated collections of programs (systems or packages).

Software may be divided into two categories: *systems software* which controls the execution of other programs and utilizes hardware effectively, and *applications software* which covers programs written to satisfy a particular user need.

Systems software is generally supplied by the hardware manufacturer. It includes operating systems; assemblers, compilers, and interpreters; programs for controlling input and output devices and copying data between storage media; and utilities for sorting, merging, and editing files, and controlling program libraries.

Operating systems are programs which enable users' software to run efficiently, making use of hardware and software resources [7.1]. They include control and allocation of processing power, storage space, and input and output devices; maintaining security; and handling files, timesharing, networking.

7.2 Programming Languages, Compilers, and Interpreters [7.2]–[7.4]

Ultimately a computer can only operate in binary machine code. Assembly language is a more convenient, mnemonic form of machine code. Its use nowadays is restricted to those applications or part-programs where another language would not be satisfactory. Third-generation, high-level languages such as FORTRAN (for scientific applications) and COBOL (for business applications) are much easier to program and to understand. Fourth-generation languages are mentioned in Section 7.6.

A written program must be compiled (i.e., converted into machine instructions) before it can be executed. A compiler sits in memory and operates as a program. The data to be compiled are known as the source program and the compiled program is called an object program.

An interpreter converts individual source statements into machine code as they are needed. Interpreters are quicker and more flexible to use than compilers but a compiler produces object programs that are independent of the compiler and execute more quickly.

One of the best known microcomputer languages, BASIC, can be either compiled or interpreted. The language C is becoming popular for microcomputers because programs written in C can be more easily made to run on a variety of machines.

7.3   Organization of Data

Data have to be organized to maximize hardware and software performance. The disposition of the data on a mass storage device is known as the physical organization but programmers are more concerned with logical organization. Items may be filed logically in sequence (a serial file) or they may be organized with special indexes used to find relevant items. Data-base management systems (see Section 7.4) can be used to provide complicated logical arrangements of data.

The nature of the physical organization may have a bearing on optimizing the software efficiency. For example, random access to items on a CD-ROM is slower than access to data on a magnetic disk, but it is then quite efficient for reading a number of items serially. Information retrieval software for CD-ROMs must be written bearing these factors in mind [6.20], [7.5].

7.4   Information Retrieval Packages

Data-base management systems and text retrieval systems are important software packages for information retrieval (see also Section 8.4.4).

Data-base management sytems are best suited to handling structured, numerical, and factual data, and short items of text, particularly if rapid updating is required [6.1], [7.6]–[7.8]. They allow different users different views of common data. Frequently they offer a high-level query language (e.g., SQL), fourth-generation programming languages, applications generators, and report generators (see Section 7.6). A data dictionary and data definition language allow careful definition of data elements and their relationships. The logical structure of the data may be hierarchical, networked, or relational. The systems are described as relational if they allow the interrelation of data elements in answering complex queries.

Text retrieval systems are best suited for large volumes of unstructured text [7.8]–[7.11]. The file structures are simpler than with data-base management systems but text retrieval systems offer specialized features such as proximity searching and thesaurus control (see Section 8.4.4).

The two types of information retrieval system tend to overlap, as data-base management systems add text-handling capabilities and text retrieval packages acquire data-base management facilities.

7.5   Microcomputer Software [6.32], [6.33]

Spreadsheets. Spreadsheet programs consist of a matrix of boxes or "cells" into which the user enters data or formulae. All spreadsheets print out data exactly as they are recorded but some also allow the production of graphs, bar charts, pie charts, etc. These programs are particularly useful for modeling "what-if" operations by changing variables and inspecting the results.

Data-Base Management. Data-base managers allow storage of data and text, and retrieval in many formats. Data bases can be created; data can be entered on data entry screens, reviewed using query commands, and reported using report generators.

Word Processing. Word processing software is used to compose text on a video display unit instead of on a typewriter. Corrections are easy to make: words, paragraphs, and blocks of text can be deleted and inserted, and spelling checker modules can be used. The final document can be printed in a number of formats.

Data Communications. This software allows communication with other computers, for example those built into laboratory instruments or the remote host computers holding scientific data bases [6.33], [7.12]. The advantages of microcomputers for uploading and downloading information in online searching of public data bases are considered in Section 8.2.7.

Graphics. Graphics software covers a wide range from simple data plotting and screen image-making programs to computer-aided design and other high-resolution design systems. Specialized software for handling chemical structures is considered in Chapter 10.

Drawing programs allow graphs and charts to be generated from data; painting programs allow manipulation of each pixel to generate images. Drawing programs are needed for analysis and communication of scientific data; painting programs are used in creative scientific communications (e.g., audiovisual aids for presentations).

Integrated Software. Integrated programs (e.g., Lotus' Symphony) combine two or more of the major program types described above.

Application environments are application programs that run under the microcomputer's operating system but provide a more advanced user interface and additional features such as concurrent operation and coresident programs. Applications programs can then be written to include code that interacts with the application environment. An example is X-windows, written at the Massachusetts Institute of Technology and adapted by various vendors. With Microsoft Windows for the IBM-PC, for example, two or more application programs can be run and information can be exchanged between applications. Windows is a graphics-oriented interface and makes the IBM-PC user interface more similar to that of the Apple Macintosh, which is characterized by the use of windows, icons, mice, and pointing devices (WIMPs).

GEM, from Digital Research, is another visually oriented user interface. It is a single-tasking environment and also uses WIMPs.

7.6   Software Engineering

The production of application programs is highly labor intensive. Developments in software always lag behind those in hardware and this can lead to a bottleneck. One solution is increased availability of packaged software. Another is the use of fourth-generation languages which allow the programmer to write instructions in something akin to natural language leaving the computer to generate code automatically.

The emergence of the new discipline of software engineering has encouraged improvements in programming practice, such as structured (modular) programming [7.13], [7.14]. Computer-aided software engineering (CASE) aims to cover the whole spectrum of software development from the technical definitions of the system to the way it is managed. The tools and methods involved include graphics, "front-end" design tools, fourth-generation languages, and object-oriented technology. Object-oriented programming systems are a way of developing packaged software that draws heavily from common experience and the manner in which real-world objects relate to each other.

7.7   Telecommunications and Networks

[6.2], [6.32], [7.15]

The function of digital networks is to interconnect computers and other devices so that data can be transferred. Online searching of remote data bases, electronic mail, and other telecommunications services are all dependent upon the transmission of digital data.

Simple data networks that connect a few pieces of equipment which are close together may simply involve lengths of cable. Access to distant computers means using the public telephone network or a private telephone line leased from a telecommunications company.

Telephone systems were originally developed for transmitting sound and are not ideal for transmission of data. The efficiency of data transmission is measured in terms of "bandwidth" (measured in Hertz), which is roughly equivalent to the maximum rate of data transmission (measured in bits per second).

Most telephone networks depend upon analogue signals whereas data consists of digital (on–off) pulses. Connection of a terminal or microcomputer to a distant computer therefore requires a modem (modulator–demodulator) to convert the digital data into analogue signals, and vice versa.

Many countries are now implementing new networks for transmitting digitized speech and data. The United Kingdom's Integrated Services Digital Network (ISDN) will gradually replace the Public Switched Telephone Network (PSTN).

In busy long-distance public networks it is important to transmit the maximum amount of data along each data line. One technique used for this is multiplexing. Data from more than one sender are interleaved for transmission by one multiplexer and unscrambled at the receiving end by another multiplexer.

Local area networks are designed to link equipment within a much smaller geographical framework. Ring networks involve a ring of cabling with each computer device connected separately into the ring. Broadcast networks (e.g., Ethernet, developed by Xerox Corporation) use coaxial cable as the transmission medium and employ separate transmitter–receivers with repeaters to boost the signals. Star and mesh networks are also in use.

International services such as the British IPSS (International Packet Switching Service) and PSS, the European Euronet, the American Tymnet and Telenet, the French Transpac, and the Canadian Datapac networks, use packet switching. This technique employs computers to control data flow and does not provide a dedicated physical path between the sender and the recipient. Messages are broken up into segments which are sent separately, maximizing bandwidth usage. The rate at which data are sent need not be fixed and devices with different data rates can communicate with each other.

Data rates on the public telephone network are limited to 48 kbits/s (often less) but local area networks can transmit data much faster (10 Mbits/s).

Wide-band digital networks rely on new technologies such as satellite links and fiber optics.

The rules governing the flow of data are called protocols. Unfortunately there is more than one common protocol. A first step towards a general-purpose network standard, allowing connection of any device to a network, is the Open System Interconnection (OSI) reference model put forward by the International Standards Organization (ISO).

7.8 Distributed Computing

The current trend is towards network computing. Within an enterprise-wide network, client/server, or distributed computing, is the predominant technology. Client/server computing splits the processing of an application between a front-end portion executing on a PC or workstation and a back-end portion running on a server. The front end typically handles local data manipulation and maintains the user interface and the back end handles data-base and other number-intensive processing. Distributed computing allows an application to run on more than one system: a PC and a mainframe, for example. Standards for data base interrogation and networking are important if applications are to operate in a "mix-and-match" fashion within a heterogenous computer environment. The Open System Interconnection (OSI) initiative started many years ago by the International Standards Organization (ISO) is significant here.

The emergence of Graphical User Interfaces (GUIs) such as Microsoft's Windows 3 and IBM's Presentation Manager have introduced the IBM user to the intuitive, WIMP-based environment already familiar to the Macintosh user. In these new environments more than one application may be open concurrently and files can be exchanged between applications either via the "clipboard" or through a technique known as Dynamic Data Exchange (DDE).

8 Records Management, Online Searching, and Information Retrieval

8.1 Introduction

Both internal (corporate) and external (public) data bases can be accessed online. To the information scientist or research chemist, the distinction between internal and external data bases is somewhat artificial. Data from an external data base can also be downloaded to make an in-house data base. In some cases the same software can be used for both the external data base and the in-house one. This chapter is concerned with information retrieval from personal, corporate, and public data bases. Chemical structure information is discussed in Chapter 10.

8.2 Online Searching

8.2.1 Introduction

In an online retrieval system a user can directly interrogate a machine-readable data base on a remote computer [8.1]–[8.6].

Online bibliographic data bases first arose in the late 1960s as byproducts of primary printed publications. As computer typesetting was introduced, more material for publication became available in machine-readable form. The creation of machine-readable data bases by the secondary information services coincided with the development of long-distance telecommunications networks, such as Tymnet and Telenet in the United States. The online industry started when two organizations with spare computing capacity, Lockheed [8.7] and SDC [8.8], provided the software and necessary computing facilities to enable the data bases to be stored and searched interactively via telecommunications networks. Nowadays many data bases are published only in machine-readable form, with no print equivalent.

Most online data bases are bibliographic in nature. The author, title, and source of the document are indexed by the data-base producer, usually with additional controlled descriptors or keywords, and possibly with an abstract. Full-text data bases are, however, also available. In addition to bibliographic data, these carry the full text of documents, footnotes, cited references, and captions for graphics or figures. At present graphics are usually omitted.

Data bases that carry factual and numeric data are described in Chapter 9. Systems permitting the retrieval of chemical structural information are described in Chapter 10.

8.2.2   Equipment

Online searching involves two-way (interactive) communication between the user and a remote computer. The user will most likely have a keyboard plus cathode ray tube display for inputting queries and receiving the output. The input device may be a "dumb" terminal, an "intelligent" terminal, or a microcomputer. For accessing systems that permit input and display of graphics, a graphics terminal or suitable microcomputer is needed. In most cases a modem converts the digital signals of the computer into the analogue signals carried by the telephone line. A modem is not required if the input device is connected directly to the remote computer or is linked to a network that allows external access. Telecommunications are discussed in Section 7.7.

The advantages of using a microcomputer are covered in Section 8.2.7.

8.2.3   Benefits and Problems

Online information systems offer rapid and convenient access to a multitude of references, facts, and chemical structures. Online information is almost always more up-to-date than that available in print. Another advantage is the number of access points. The indexes to a printed volume limit the number of ways in which information can be accessed. Moreover, in searching online the user can combine terms using Boolean logic, for example:

A or B or C or D
A and B and C and D
(A and B) or (C and D) not (E and F)

Keywords, authors, formulae, patent numbers, chemical names, and many other fields, can be searched in one or more data bases. In some systems one query can be simultaneously submitted to more than one data base. "Cross-file" searching amongst certain data bases is also possible (i.e., carry a cross file of reference numbers selected in one query and search for records bearing those reference numbers in another data base or system).

Considerable expertise is needed to carry out all but the simplest of online searches. For many years almost all searches were carried out by an information scientist or skilled intermediary. In recent years there has been an increasing tendency for users (scientists, lawyers, or managers) to do the search themselves. End user searching has many advantages [8.9]. End users have immediate access to information relating to their problems. Their flow of ideas is not interrupted and searching can be performed interactively (e.g., by modifying the original query). Use of subject expertise in the search formulation is increased. The end user may be encouraged to use systems and to do searches when the intermediary is not available. Users become more skilled at explaining their requirements if they use an intermediary on occasions. End-user searching allows an information unit to concentrate its expertise on more complex problems. In organizations with few (or no) skilled intermediaries, end users are obliged to do their own searching, or to pay an information center or broker.

The average online system is not very user-friendly [8.10]. Even if the user knows which data base to consult and the host, the logon procedure is laborious. The host's command languages and error messages are not user-friendly. Indeed proper use of such software requires extensive training and voluminous documentation. Each vendor supplies separate contracts and passwords. Customer service is not available around the clock. Document delivery mechanisms are cumbersome. Data-base names vary from system to system. (In 1985 there were 17 names for the various CAS files on six different computer systems.) Commands also vary from system to system. Attempts to standardize on one command language [8.11] have made little impact on the average user. Online Inc. of Weston, Connecticut produce an International Command Chart comparing the command languages of various systems.

End users take longer than an intermediary to run a search [8.12]. Some systems designed for end users do not have all of the searching capabilities of the traditional systems. End users have difficulty interpreting output and revising an ineffectual search strategy; they are also unaware of or do not use search aid tools such as thesauri. Some users do not understand Boolean logic; many do not use very complex search strategies. End users are infrequent users and therefore forget commands and search protocol.

A reasonable compromise in large commercial organizations is for end users to do the "quick-and-easy" searches while experts perform complex searches, or searches on which major financial decisions might depend.

The development of tools to facilitate end user searching is considered in Section 8.2.7 and Chapter 10.

8.2.4   Data-Base Producers and Vendors

The company (or system) which has the hardware and software that makes a group of data bases available is known as a *host* or *vendor*. Some American sources refer to hosts as *data banks*, whereas many European sources use the word data bank for a data base of factual and numerical information. A company which constructs a data base is known as a *data-base producer*. Some data-base producers are also the vendors of their own (and other) data bases.

About 4500 online data bases are now available (1990) on about 600 different hosts [8.13], [8.14]. Directories are available in hard copy and online. Major hosts are:

Dialog Information Services, Palo Alto, California
Maxwell Online, McLean, Virginia (incorporating the older services BRS, SDC/ORBIT, and Pergamon Infoline)
STN International, Columbus, Ohio (see Section 2.2.6)
Questel (Télésystèmes), Paris
Mead Data Central of Dayton, Ohio
European Space Agency Information Retrieval Service (ESA/IRS), Frascati, Italy
DataStar, Berne, Switzerland
Deutsches Institut für Medizinische Dokumentation und Information (DIMDI), Köln, FRG.

The preeminent chemical data base is that produced by CAS (see Chap. 2). Chemical structure searching is possible on STN and Questel. Versions of the CAS data base are available on various hosts but only STN International offers searchable abstracts online and pre-1967 registry data.

If vendors do not offer chemical structure searching, compounds have to be located by chemical names, registry numbers, or molecular formulae. Often the vendor supplies a chemical dictionary file where the user may find the preferred name or a registry number for a compound. The preferred name, or the reference number, can then be used for access to data in another file.

Several chemical hazards bibliographic data bases are available [8.15].

SCISEARCH, the online version of Science Citation Index (see Section 2.3.1) is available on more than one host. It is not a bibliographic data base in the same sense as others listed here, but it is useful to include it here. SYNGE has reported on its use as a complement to Chemical Abstracts [8.16].

In the patent information area, Derwent Publications' World Patents Index (WPI) and WPI/L (where "L" stands for "latest") are of great importance. Searching generic structures is possible (see Sections 20.3 and 10.4.1). IFI Plenum's CLAIMS files cover U.S. patents and have chemical data back to 1950. APIPAT and APILIT are produced by the American Petroleum Institute. The Chemical Abstracts data base also covers patents. MARPAT is STN International's service for searching generic structures from patents (see Sections 10.4.1 and 20.2). The French patent office, INPI, has data on Questel. INPADOC is the most comprehensive collection of worldwide patent literature and it builds up and reports the various Patent Family Collections. Mead Data Central offers the full text of U.S. patents as LEXPAT (this is not simply a bibliographic data base, but is mentioned for completeness).

Derwent Publications' Chemical Reactions Document Service (CRDS) is a rather complex system based on various codes (see Section 3.4).

Other data bases are mentioned in Chapters 2 and 3. For a fuller listing of data bases in chemistry and related disciplines, see [8.13], [8.14], [8.17]–[8.20].

Data banks (i.e., files of factual and numeric data) of interest to chemists are considered in Chapter 9. Beilstein Online (see Sections 3.3.2 and 9.6.1) is one of the most important data banks.

8.2.5   System Features and Search Strategies

A checklist of the functions and capabilities of the systems of many major vendors has been compiled [8.21]:

Access to system
Data bases mounted
Data base selection
Treatment of records
Searching capabilities
Printing offline
Limits on capacity
Saving searches
Multiple file and cross-file searching
Searching assistance while online
Vocabulary assistance during a search
Simplified searching options
Information on costs and system usage
Online ordering capability
Documentation available from vendor
Training available from vendor
Vendor charges
Electronic messaging
Software package enhancements

Online bibliographic data bases are based on the *inverted file structure*. Typically, an inverted file structure consists of several related files. The "main" or "record" file stores the bibliographic records themselves. These records may be maintained in any order, usually the order in which they are acquired. To search a large data base for all the occurrences of one term would be very time-consuming if this file were treated as a "direct file" and searched sequentially from beginning to end. To speed up the search process, and facilitate more complex searches for combinations of terms, an index file is created. The index is usually stored and used in a two-step operation.

The main index stores a list, typically in alphabetical order, of all the terms in the records file together with the number of times that term occurs in the file. A secondary index stores the record numbers where each term is found. When a search is initiated, the software begins by searching the main index file and if the search term is found reports the number of occurrences (postings) to the user. If the user wishes to display the results, the secondary index (postings) file is accessed to obtain the record numbers in question. Only at this stage does the software actually go to the records file and pull out the correct text identified by the record number.

Searching using an inverted data-base structure is therefore very fast and efficient because in the first instance only the main index need be searched. Data-base vendors have introduced their own refinements to this basic structure.

In addition to the use of Boolean operators, word stem searching (truncation: e.g., CRYST* to find crystal, crystallography, crystallographic, etc.), proximity operators, and other facilities may be available.

Formulation of a search is not simple. A strategy should be developed on paper before costs are incurred formulating the search online [8.19]. Detailed examples of search formulation for the Chemical Abstracts data base are given in [8.18], [8.22]. A complete search may require the use of more than one data base.

The complementary nature of the SCISEARCH and Chemical Abstracts data bases is described in [8.16].

8.2.6   Costs

Data-base producers receive royalties from online vendors and in some cases charge customers subscriptions. Often, online searchers receive a discount on online costs if they subscribe to the hard-copy version of a machine-readable data base.

Online vendors used to base their charging heavily on the number of minutes for which the user was online (connect charges). However, due to development of offline search formulation and rapid searching using fast modems, vendors introduced other charging systems. Charges for opening a data base, for each search term used, for each hit displayed, or for each offline print ordered are all used [8.23].

8.2.7   Gateways, Front Ends, and Microcomputer Software
[8.24]–[8.31]

Definitions. The problems described in Section 8.2.3 have been addressed by software packages, loosely called gateways, front ends, and intelligent interfaces, many of them operating on the user's microcomputer. The microcomputer has the advantage of allowing the user to do online searching on the same "terminal" used for other tasks. The user thus has an ever-increasing volume of storage space, can make use of specialized searching and utility software, and can keep statistics on performed searches.

Gateways are systems that allow the user to switch easily from one host computer to another, often with a simplification of invoicing procedures, and sometimes permitting the use of just one command language for more than one host system. Terminal emulation software (sometimes also called a communications package) [8.29], [8.30], [8.32]–[8.34] is required so that the remote computer regards the user's personal computer as a terminal and can communicate with it. Sophisticated communications software offers additional features such as automated logging-on and the ability to capture data output from the remote computer on disk (downloading).

Front ends [8.10], [8.34]–[8.38] offer many more features, including some at the "back end" of the search:

1) Access to several hosts
2) Data base selection
3) Automatic dialing and logon, including the pursuit of an alternative if the chosen route is unobtainable, and the handling of error messages

4) Offline search formulation and storage of profiles
5) Presearch editing and uploading (allowing interaction and uploading of selected statements)
6) Online running of the search when required, including facilities to cope with error messages
7) Help features: there should be a different question-and-answer dialogue, or menu-driven interface, for novices from that needed by expert users
8) Viewing and printing of results
9) Transfer to a different data base
10) Downloading
11) Logoff
12) Reformatting and post-processing of the search results, with more powerful features than those available in word-processing and other utility software

Search Formulation. The user interface is usually menu-driven or works in question-and-answer mode. Menus are essential for the naïve and frustrating for the experienced: a good system offers both user and expert modes. Natural language interfaces and other artificial intelligence techniques to facilitate searching for the novice (e.g., TOME Searcher) are described in Section 11.5.2. Verity's TOPIC [8.39]–[8.41] is a concept-based retrieval system for use in distributed computing environments. The Intelligent Test Management System is an expert system from Information Access Systems [8.40]. For a historical review of earlier systems, see [8.42].

Uploading [8.31]. Formulation of a search offline saves money (by eliminating connect time) and reduces the stress on the user (the "taximeter syndrome": the urgency of formulating the search quickly, and perhaps carelessly, because money is being spent by the minute). The user can then log on to a remote computer and "upload" the query.

Storage of search profiles and uploading is useful in the following cases [8.31]:

1) For the creation, rapid transmission, and re-use of "hedges" (i.e., the set of terms that defines a common or regularly used search concept)
2) For a search to be used against several data bases
3) For a search that is run daily or weekly
4) For a "boiler-plate" search, where only a few terms, amongst many, are changed at each execution of the search
5) For a "canned" search (e.g., one which is stored to run at 3 a.m. at cheap telecommunications rates)

Downloading [8.34], [8.43]. The major benefits of using a microcomputer for online searching were first thought to be automatic logon and offline search strategy preparation. Improved storage technology, faster telecommunications, and the availability of suitable software made downloading onto disks possible. It is both faster and cheaper to omit the printer, and multiple copies of the search output may be made from the disk. However, the most important benefits of downloading are the possibilities of manipulating the data and producing in-house data bases or bulletins. Word-processing or data-base management software can be used for this but specialized packages (see Section 8.3.1) offer advantages [8.25], [8.34], [8.44]–[8.46].

Other Search Aids. Some vendors offer simplified systems on the host computer for the benefit of end users. Examples are BRS BRKTHRU on Maxwell Online and DIALOG's Knowledge Index.

"Gateway services" such as Infotap's Intelligent Information and Istel's Infosearch [8.47] also enable the searcher to access data bases on different hosts via one logon procedure and password. Some gateway services, such as EASYNET [8.48], [8.49], also offer help in choosing a data base and formulating a search strategy.

8.2.8   Full-Text Online Data Bases [8.50]

Full-text data bases comprise the complete texts of articles, books, newspapers, encyclopedias and so on, available in a machine-readable form in which every word in the entire text (except designated stopwords such as "the" and "and") can be searched.

In 1988 DIALOG offered the entire text of 355 periodicals [8.51]. Mead Data Central's NEXIS system offers the full text of many magazines and newspapers [8.50]. Mead also offers the text of U.S. patents in LEXPAT. STN International's Chemical Journals Online (CJO) [8.52] is described in Section 1.8. The full text of the Kirk–Othmer Encyclopedia of Chemical Technology [8.53] (see Section 3.3.1) is available online [8.54].

Most full-text data bases are searched with the same sort of software as is used for searching bibliographic data bases. This is rather unsatisfactory because with full-text data bases the context of the search terms is very important. In a search for (A AND B) false drops will occur if the terms A and B are separated by ten pages of text. Use of Boolean operators plus term truncation and synonym searching is not sufficient. Proximity operators (e.g., DIALOG's proximity operator "S") are needed to ensure that search terms occur in the same chapter, paragraph, or sentence.

A further enhancement to full-text searching would be the use of a term count feature. Information retrieval systems usually report the total number of occurrences (or postings) of a term in the whole data base. The greater the number of occurrences, the greater the relevance of the document.

Synonym listings, ranking, term weighting, and document clustering [8.55] are other techniques that can enhance retrieval [8.56]. These are discussed in Section 8.4.4.

Specialized techniques are needed for the construction of full-text data bases [8.57], [8.58]. The use of a generalized markup format for electronic publishing of a hard-copy journal can facilitate conversion of the data into an online full-text data base.

Large backfiles of journals are not available and end users are not yet avid searchers. Unfortunately graphics, mathematical symbols, and tabular material are not usually mounted online because of the expense. This reduces the usefulness of scientific journals online. More advanced electronic publishing systems will facilitate the inclusion of graphics in full-text journals online [8.59].

8.2.9 Graphics Display

On STN International PATGRAPH is the patents graphics subfile of the German patent data base PATDPA. It contains patent drawings, chemical structures, and complex mathematical formulae published in addition to the abstract of German first patent publications (see also Section 21.3).

An international facsimile standard called "Group 4" has been established by the Comité Consultatif Internationale Télégraphique et Téléphonique (CCITT). Graphic data for PATDPA are scanned in this format and converted to outline vector representation (see Section 7.2.2) using a system called SCORE (Scan Conversion for Outline Representation of Images) developed by Imagin, Karlsruhe, F.R. of Germany [8.60]–[8.64]. The transmission of text, patent drawings, images, and chemical structures via telecommunications networks is possible since text and vectorized graphics are transmitted as ASCII data (see Section 6.1). Transmission of vectorized data is much more effective than transmission of the original digitized graphics: the compression factor is more than 60%. Vectorization and the appropriate software (STN Express, see Section 10.6.4) facilitate the display and printing of graphics on a wide range of graphics screens and printers.

The DIALOG data bases TRADEMARK-SCAN-FEDERAL and HEILBRON contain graphics which can be displayed (but not searched) with DIALOGLINK software [8.65], [8.66]. Macintosh microcomputer users can print them using DIALOG Image Catcher. TRADEMARKSCAN-FEDERAL contains all active trademark applications and registrations filed in the United States Patent and Trademark Office. Trademark images have been digitized and compressed. HEILBRON is the online version of the Dictionary of Organic Compounds, DOC5 (see Section 3.3.3) [8.67]. It contains scanned chemical structures.

A more recent development in DOC5 is the "OCR-ing" of chemical graphics [8.68] (for a description of optical character recognition, OCR, see Section 11.5.2). From Chapman and Hall's Dictionary of Organic Compounds, 55000 chemical structure diagrams have been scanned by a new program, Recog, to produce parameterized diagrams which can be converted by a further program, Constr, into a substructure-searchable data base. This technology opens up new possibilities for producing structure-searchable data bases from hard copy and is rather different from the usual concept of drawing structures and publishing with the same structure drawing package (see Chap. 10). IBM are also studying OCR of chemical structures.

8.2.10 Electronic Document Delivery [8.69]–[8.72]

Improvements in document delivery were alluded to in Section 1.5. Electronic mail is a comparatively recent innovation as far as online searching is concerned. Several hosts have had some form of electronic mailbox system for online document ordering for some time. It was not until 1988, however, that the integration of electronic messaging, facsimile (FAX), and telex with searching facilities became available. A customer can now immediately download a search into his mailbox and deliver the result to the requestor's mailbox or convert it into a fax or telex.

The British Library project, QUARTET, is researching electronic document delivery [8.69], [8.70]. They envisage a "just in time" concept for information. The user pays only for what is needed. Users find articles they want or have the bibliography checked by online searching, request information from the British Library by electronic mail, and receive documents by Group 4 FAX over ISDN. The Group 4 FAX standard has been incorporated into the new international standards that allow compound documents to be transmitted between various devices. QUARTET makes use of the ADONIS (Article Delivery Over Network Information Systems) CD-ROM data base of articles in biomedical journals [8.71].

8.2.11   Videotex [8.73]–[8.79]

Videotex is synonymous with *viewdata* (the latter term is usually used in the United Kingdom). Both terms are generic and include any interactive systems for transmitting text or graphics stored in computer data bases via telecommunications networks for display on television screens. Color is also part of the standard. The classic example of viewdata in the United Kingdom is the PRESTEL system.

Videotex should not be confused with *teletext*. Information from teletext systems is also displayed on a television screen but forms part of the regular broadcast transmission. Teletext is noninteractive. Two examples of teletext systems are Ceefax and Oracle in the United Kingdom.

Videotex systems can be accessed using special terminals, modified television sets, or microcomputers with videotex software. Standard asynchronous communications software cannot be used.

The approach to data bases is different from that for bibliographic systems because videotex systems must be cheap and very easy to use since they are designed for use by large numbers of the general public. The data-base structures and information retrieval techniques have to be designed accordingly.

Data are stored in tree structures. The nodes of each tree are called pages and they are identified by page numbers related to the position of the page in the tree. Pages are divided into frames. The frame numbers are differentiated by a letter following the page number.

Videotex systems are menu-driven. The user can browse through increasingly specific frames of information or access a page directly. Page numbers may be located through the printed indexes or index pages online. Good menu design and up-to-date indexes are essential.

The use of tree structures in a menu-driven system is simple but limits information retrieval facilities. Some information providers have therefore developed better types of index plus limited keyword searching.

Videotex systems can also be set up in-house or by using a bureau. Videotex has proved very useful in the travel industry, stock control, sales records, pharmaceutical wholesaling and distributing, clinical trial management, adverse drug reaction reporting, and in communications with sales representatives in the field.

European videotex systems which were created in the 1970s and early 1980s were set up on a national basis. Each country had its own policy for technical standards, networks, terminal distribution, subsidy to information providers and tarification. The general objective, however, of all systems was to provide a medium as cheap and simple as the telephone for distributing quickly changing information (e.g., public transport timetables and directories).

France, with its distribution policy of free or cheap terminals, had the largest installed European base of 5.2×10^6 terminals by February 1990. The Federal Republic of Germany and the United Kingdom had only 163 000 and 155 000 terminals, respectively.

8.3   Public Data Bases For In-House Use

8.3.1   Microcomputer Data Bases

In-house data bases can be purchased (on tape, floppy disk, or CD-ROM) or can be constructed in-house from downloaded data, copyright permitting.

Downloading is time-consuming and expensive. Hit files from various searches and hosts need to be merged and duplicates have to be detected and removed. The search software for the in-house data base may be different from the hosts' information retrieval packages. Display features differ. File inversions on the microcomputer may well tie it up for hours. However, as storage capacity, the speed of PCs, and software improve, many more users will start constructing data bases from downloaded data.

Much software is now available for downloading data and making personal data bases. Examples are Pro-Cite with Biblio-Links, Reference Manager with Capture, Bib/Search, VCH, Biblio, and InMagic with Headform [8.25], [8.34], [8.44]–[8.46]. Many of the major text retrieval packages such as STATUS, CAIRS, and BASIS are available in microcomputer versions [8.80].

Some information providers sell data on floppy disks, e.g., Current Contents from ISI (see Section 2.5). The greater capacity of CD-ROM has, however, made it a more suitable medium for the supply of information for use in-house.

8.3.2   Data Bases on CD-ROM

CD-ROM as a storage medium is discussed in Section 6.3. Its large capacity and the fact that it is a read-only medium make it suitable for supplying full-text public data bases for in-house use. It also has the advantage that the full "text" can contain graphics.

There has been much controversy as to whether CD-ROMs can compete financially with online data bases [8.81]. In searching a CD-ROM the user does not have to worry about telecommunications problems or connect charges. The retrieval software is more user-friendly than that of online systems (it is often hypertext-like, see Section 11.7), although it may not be as fast or powerful. Unfortunately CD-ROMs cannot be updated as fast as online data bases, consequently CD-ROM is most suitable where currency is not essential but data base access is very frequent. Enormous data bases, such as Chemical Abstracts, are too large to be supplied on CD-ROM because too many discs would be needed.

The uses of CD-ROM are described in [8.82]–[8.85]. Directories of available data bases have also been published [8.13], [8.86]–[8.88].

DIALOG offer a number of their online data bases on CD-ROM as DIALOG OnDisc Products. The National Library of Medicine's MEDLINE is one of these. CSA Compact Cambridge (a Cambridge Scientific Abstracts service) also offer MEDLINE, as do SilverPlatter of Wellesley MA. Compact Cambridge produce many CD-ROMs for the medical community, including Drug Information Source, produced in collaboration with the American Society of Hospital Pharmacists.

The EMBASE data base of Excerpta Medica in Amsterdam is marketed on CD-ROM by SilverPlatter.

The MICROMEDEX TOMES Plus System (Micro-info of Alton, UK) offers the Registry of Toxic Effects of Chemical Substances (RTECS) and other health and safety data. OSH-ROM (SilverPlatter) contains three health and safety data bases, including RTECS.

BIOSIS's Biological Abstracts on CD-ROM are available through SilverPlatter.

CD-ROMs of interest to the agriculture industry include the Royal Society of Chemistry's Pesticides Disc, produced in collaboration with Maxwell Communication Corporation; the UN Food and Agriculture Organization's AGRIS Data, marketed by SilverPlatter; and CAB Abstracts, from CAB International, also marketed by SilverPlatter.

ISI sell a CD-ROM version of Science Citation Index.

The ChemLink Fine Chemicals Data Base (text only), produced by Chemron of San Antonio, TX, lists sources for 137 000 chemicals. The CD-ROM version is sold by Microinfo. (An online version is up on DIALOG.) The Aldrich Chemical Company produces its catalog on CD-ROM.

The European Patent Office's ESPACE system, with software written by JOUVE, has text and graphics for recent European and PCT patents on multiple CD-ROMs. Chadwyck Healey offer the text of U.S. patents since 1973 but there are no graphics.

Whitakers Books in Print and Bowkers Books in Print are both available on CD-ROM. IBM and Oxford University Press collaborated in the production of the Oxford English dictionary on CD-ROM. A more recent product is Molecular Structures in Biology, a reference tool for exploring protein structures [8.89]. The software supplied allows both molecular graphics options and special hypertext facilities.

Current Facts in Chemistry on CD-ROM contains the most recent year's worth of data from the Beilstein Institute. Numeric, factual, and substructural data can be searched, the last by means of S4 software (see Section 10.5.9).

8.4 Records Management [8.90]–[8.92]

Most organizations have large amounts of proprietary data (reports, correspondence, laboratory notebooks, etc.). Policies and procedures need to be established for managing such records: determining what should be stored and how, establishing retention times, assigning security categories, controlling archives, and adhering to standards required by regulatory authorities.

8.4.1 Microforms [8.93], [8.94]

There are two principle microform products, flat microfiche and roll microfilm.

Roll microfilm, usually 16 mm, is the preferred medium for purely archival microfilming; 3000–10 000 documents can be stored per film. Roll film can be optically marked (with a "blip") or coded to provide a degree of automated retrieval. If a computerized index to the documents is held they do not have to be sorted prior to filming.

Microfiche is an alternative to film but is more expensive to produce. Fiche is preferred to film if a number of copies are to be made and higher quality is needed. Standard microfiche carry 98 frames but 60–420 frames are also used.

An updatable version of microfiche is the *microfilm jacket*, where strips of 16-mm microfilm are inserted into transparent pockets on a standard fiche. Updating such systems is labor-intensive but can save retrieval time if the contents of a single file are not scattered across multiple films.

Computer output can be linked directly to the microfilming process, thus eliminating the need for an intermediate paper copy. *Computer output microfilm* (COM) may be generated in-house, or the data may be spun onto a magnetic tape which is sent to an outside bureau for COM production, usually on fiche rather than film.

Major vendors in the micrographics area are Kodak, Bell and Howell, 3M, and Agfa.

As a storage medium, microforms offer great space saving (98% compared to paper) and stability (with an archival life of up to 100 years). They can be easily and speedily produced, duplicated, and transported. Microfilm can be generated directly from computer files, can be accessed by computerized indexes (with some difficulty), and copes with a variety of textual and graphical material.

Many users dislike reading from film rather than paper but the main problem is the slow, two-step retrieval of information. Identification of the appropriate roll or fiche and the location of the relevant frame can be automated (computer-assisted retrieval, CAR). For roll film attention has been focused on automatically locating the frame after manual selection of the roll. With fiche more attention has been paid to automated access to the relevant fiche with manual location of the desired frame. Computers allow automatic prefiling and random retrieval on fields such as account number or date, rather than roll and frame numbers. CAR systems are usually dependent on key entry for indexing purposes but many organizations deploy equipment with integral bar code readers that can automatically index and capture the document image in one pass.

Kodak's KAR, Agfa's ADMIS (Agfa Document Management Information System), and Bell and Howell's DP1000 are well-known CAR systems. In Kodak's KIMS the retrieved microfilmed image is scanned electronically and converted into a digital bit stream for transmission to a remote terminal. The image can be viewed or processed further.

8.4.2   Document Image Processing [8.93], [8.95]–[8.97]

Document Image Processing (DIP) is the technology of storing and retrieving documents from WORM (Write Once Read Many) optical discs (see Section 6.3). These systems are also called optical filing systems. A 12-inch disk can hold up to 2 Gbytes of data (ca. 50 000 A4 pages).

Basic DIP systems provide fast access and retrieval (10–30 s), concurrent access by several users, indexing and availability for retrieval of new documents within hours of being received, enormous storage space savings compared with paper, and remote access by a PC over a dial-up telephone line with automatic delivery by facsimile transmission.

The constituent technologies include scanners, optical storage (discs, drives, and "jukebox" selectors), OCR, indexing and text retrieval software, and facsimile [8.98]–[8.102].

Documents are scanned using a facsimile-type scanner. The images scanned in raster form are indexed, compressed, and stored on an optical disc as a digital bit map. Retrieval consists of carrying out a search using an index system, and passing the request via a document manager unit (which may be connected also to a temporary magnetic storage unit) to the optical disc. The retrieved image is displayed on a high-resolution screen. Documents can be output to a laser printer.

Vendors of DIP include Wang (WIIS), FileNet (FileNet and British Olivetti), CACL (Intelligent Archive), Ingenium Software (Archea), Philips (Megadoc), Callhaven (MARS), Kodak (KIMS), IBM (ImagePlus), Document Systems Limited (InfoPlus), Image System Integrators (Image System), and Xionics (DIP-X). The micrographics companies Kodak, 3M, and Bell and Howell are involved in both CAR and DIP. With Kodak's KIMS 4000 system, users can scan documents and store them on optical disc, then download the index data to an IBM PS/2 or IBM mainframe computer. Multiple users can access a document index data base and search for documents.

One use for DIP in the pharmaceutical industry is the filing of a New Drugs Application (NDA). Unfortunately the legal status of optical disc files has not been established and companies are obliged to retain paper copies of all the data they file optically. Microfilm has legal status in most countries.

The European Patent Office (EPO) has scanned a huge backfile of patents into raster image format and would also like to accept all new patent applications in electronic form. The text must be held in character-coded form and the image data in raster or vector form. Patent applications can be filed on floppy disk or paper. The EPO uses a modified version of ISO 8879 (Standard Generalized Mark-up Language, SGML) to mark-up patents for entry into full-text data bases [8.58], [8.103]. The text can be edited and printed as required.

There is thus a dual system. For backfile conversion CCITT Group 4 facsimile standards were followed (see Section 8.2.9) and the patents were stored as images. This means that they are searchable only by the keywords used to index them and not by content. The patent text cannot be edited. For current needs, the patent text was captured in character-coded form and marked-up for entry into full-text data bases. It can thus be searched, edited, and printed.

The storage of the average A4 page on an imaging system occupies ca. 30 times more space than that required by normal character codes. As most business documents consist of a large amount of text, total image capture is a wasteful method.

New, standard document architectures will eventually allow image documents to be broken down into image and coded components for storage and distribution, and rebuilt in the workstation. Line loadings and storage space requirements will then be very much less and the current need to separate image system networks from data networks may disappear. Cheaper hardware with greater functionality will allow separation of documents into the required architectural components. Higher speed lines will aid distribution of documents.

Two developing international document architecture standards are Office Document Architecture (ODA) and Office Document Interchange Format (ODIF) under the Open System Interconnection (OSI) model. Computing companies have their own standards: for example IBM's Document Content Architecture (DCA)/Document Interchange Format (DIF), and DEC's Compound Document Architecture (CDA)/Digital Document Interchange Format (DDIF) [8.104].

8.4.3   Comparison of Micrographics and Optical Filing [8.105]

Optical filing allows huge storage capacity on a single disc, with jukeboxes allowing access to up to 200 discs. Large collections of documents and drawings can be held and retrieved automatically in seconds. Once character-coded data can be integrated, there will be further advantages. Retrieval from microforms is slower and less convenient.

Micrographics is a well-established technology whereas DIP is much newer, lacking in some standards, and very expensive. Microfilm is stable for up to 100 years: a life of 10 years is claimed for optical discs. The legal status of optically stored documents is in doubt.

An image on an optical disc can be accessed by any user connected to the system. Few micrographics systems allow remote access by multiple users over telecommunication lines. Multiple users are commonly accommodated by multiple microfiche copies.

All the user of an DIP system requires to access the images is a single workstation; there is no need for both a computer terminal and an image terminal as with CAR systems. Finally, DIP systems offer image enhancement facilities for poor quality documents.

8.4.4   Information Retrieval [8.25], [8.34],[8.56], [8.80], [8.106]

Types of Data Base. A data base differs from a conventional file in that the user expects to be able to access any piece of data in more than one way.

Multi-indexed data bases have the inverted file structure described in Section 8.2.5. The records are stored in one file and accessed by any key included in the main index. This type of data base is very easy for the user to create and maintain. Entries are made in the main file and the system automatically does the indexing. Only records of the same type can be stored.

A *hierarchic data base structure* can store records of different types and works on the principle of sets of information and the relationships between them. A data base of sales orders, for example, may contain records relating to orders, products, parts of products, and customers. Customers will order parts, products may have many parts, orders will have dates, etc. There may be one-to-one, one-to-many, or many-to-one relationships between the various items. In a hierarchic data base all the records are linked together by means of pointers so that relationship can be identified. Setting up the data structures is complicated.

A *relational data base* has very complicated data structures. The principle is again based on the relationship between records but the data base can relate records from different files. Records are linked and cross-referenced by the contents of fields that are common to the different files.

Data-Base Software. Simple filing systems are designed to emulate card index systems and information retrieval is usually only possible using one or two keys designated when the system is set up.

Data-base management systems and text retrieval software were compared in Section 7.4. Text retrieval software is designed to cope with large volumes of unstructured text. It is generally easier to use than a data-base management system but, because it is concerned more with relevance and recall than with precision, it has fuller retrieval facilities. These are similar to those found in major online systems, e.g., Boolean searching, term truncation, field searching, and proximity searching.

In contrast to a data-base management system, records are usually stored sequentially. The inverted index model is likely to be used. Report generation and sorting facilities are usually available.

Unlike a data-base management system, concurrency is low, i.e, it is not usually possible to use the same data for two different activities at the same time.

Many text retrieval packages are on the market [8.80]. Some are used by major online vendors, e.g., DataStar uses BASIS, which is marketed by Information Dimensions. Other well-known packages are TRIP (Paralog), ASSASSIN (Associated Knowledge Systems), and BRS/Search (BRS/Search).

There are also many information retrieval packages for use on microcomputers [8.25], [8.34], [8.80].

Future Developments. Work carried out in the 1970s and 1980s has been aimed at producing a new type of document retrieval system which is both more effective, in terms of the relevancy of information retrieved, and more efficient in terms of human effort.

Manual indexing is time-consuming, expensive, and prone to error: automatic indexing could offer advantages [8.107]–[8.111]. It is possible to identify potential indexing terms (ignoring stop words such as "and" and "the") in the machine-readable titles and abstracts of the documents to be searched. These content words are then passed to a stemming algorithm which finds the word root. For example, "absorbed" and "absorption" would both reduce to "absorb".

In work pioneered by SALTON [8.112]–[8.114] a document in a collection is described in terms of weighted vectors, a set of numbers representing importance-weighted concepts. Queries are assigned vectors in the same way and the collection is searched by matching document and query vectors. Retrieved documents are reported in rank order, queries are modified (relevance feedback) and the search is repeated until the user is satisfied. SPARCK–JONES [8.107], [8.108], [8.115], CLEVER-DON [8.116] and WILLETT [8.56] have also published extensively in this area.

The problems involved in formulating a search using Boolean logic can be avoided if a best match or nearest neighbor search algorithm is used to rank the documents in a collection in order of decreasing similarity with the query [8.56], [8.116]. Similarity is measured using the weights assigned to search terms. The documents output as hits can be inspected in order of decreasing similarity, giving the user control over the size of output.

WILLETT has studied the use of parallel processing (see Section 6.1) in document clustering [8.55], [8.117]. He has also compared statistically-based techniques of information retrieval with knowledge-based ones. Expert systems and WILLETT's comparison are considered in Section 11.6.7.

The recent development of hypertext is detailed in Section 11.7.

9 Numeric and Factual Data Bases (Data Banks)

9.1 Introduction

This chapter describes publicly available collections of numeric, textual, and factual data relating to chemical compounds. In Europe, such data bases are known as data banks. Chemistry is particularly well served by the bibliographic and chemical structure data bases (Chaps. 8 and 10). The compilation of numeric and factual data bases is equally important because they allow systematic analysis of large numbers of related structures and associated data; however, the development of these data banks has been less well coordinated. A large number and variety of data banks are available online to users. Data can cover several different properties of a chemical compound, but most data banks assemble facts relating to one category of data (e.g., NMR spectra) in such a way that the user does not need to access the original information source. For a data bank to be of value, the data must be correct and validated.

9.2 Types of Data Banks and Data

Data banks in chemistry may be broadly classified as follows:

1) Spectral data
2) X-ray crystallographic data
3) Chemical and physical properties
4) Toxicology and biological properties
5) Environmental and hazard data
6) Data relating to legislation

Many data banks contain data on more than one of these categories. Compilation of a large, high-quality data bank demands motivation, justification, funding, coordination, and organization. The impetus behind the creation of a data bank has tended to come either as a result of legislation or in response to a need by a group of users. Because of the limited space available in the primary literature, it is not always possible to include detailed experimental results, and data banks have been created as a depository for complete experimental results.

Chemical data banks contain a variety of data types: structural, textual, numerical, and graphical. Software for handling bibliographic data has been in use since the early days of machine-readable data base preparation, but chemical structures require specialized software. Much research has been devoted to this topic [9.1]–[9.8], with the result that the software for storage, retrieval, and manipulation of chemical structural information is probably more advanced than that available for handling most other data types.

Textual data is used in nearly all data banks, e.g., for recording chemical names and synonyms, or for chemical hazards and action to be taken in cases of emergency. Emergency information must be up-to-date and accurate.

Numerical data, coordinates, and codes present different problems. The storing of numerical data is easier than text, but facts should be validated and correct. Specialized search techniques are needed. For example, the user may wish to search for a range of values and combine the data search with a chemical structural search. Three-dimensional structure coordinates require graphics hardware and software.

An integrated approach to numeric, textual, bibliographic, and graphical information is required so that data can be transferred between files, or data banks created that incorporate all three types of information. Integrated systems for storage, manipulation, retrieval, and printing of structures and text are less well advanced, but are required for hard-copy versions of data banks, such as DOC5 [9.9].

9.3 Quality Control

The input of accurate, validated data is of paramount importance in the creation of any data bank, but particularly so where the data is used to assess the risk to humans and the environment [9.10]. Data may be used for making major environmental policy decisions at national and international levels, or for calculations and predictions.

Just because data is available online, it is not necessarily accurate because it may not have been evaluated. Errors already in a data bank are difficult to remove. They are best prevented at source. Transcription errors can be avoided by taking the data directly from the primary source. Few organizations can afford the high-cost, labor-intensive methodology used for evaluating data in the Beilstein Handbook [9.11]. Regrettably, high accuracy was offset by poor currency, and a new philosophy has been adopted for the online data base still with high quality control standards [9.12] (see Section 9.6.1).

The Standard Reference Data (SRD) program at the National Institute of Standards and Technology (NIST) improves data quality by supporting ongoing data centers; funding short-term data evaluation projects; and cooperating with other groups in the government and private sectors [9.13].

The Cambridge Crystallographic Database (see Section 9.5) also contains highly accurate, quality-controlled data. Bond lengths are calculated from the published cell and atomic coordinates and are compared with the bond lengths quoted by the author [9.14].

Data evaluation and quality control are more difficult in the field of toxicology. The Registry of Toxic Effects of Chemical Substances (RTECS) data bank (Section 9.7), compiled and updated by the National Institute for Occupational Safety and Health (NIOSH), only uses data from refereed journals.

Data accuracy has been studied in the development of the Pesticides Properties Data Base at the USDA Agricultural Research Service (ARS) [9.15]. Few bibliographic abstracts contained the required data, and few hard-copy data sources contained literature citations. A consistent and objective data evaluation scheme is being developed, together with a thorough report of experimental procedures and conditions.

Some of the factors chemists need to consider in determining the reliability of nonevaluated data are listed in [9.16]. The quality of spectral data bases is discussed by SHELLEY who is concerned about errors, omissions, lack of diversity of structures, and the problems of integrating spectral and chemical structural search systems [9.17].

The two main commercially-available mass spectral data bases, the Wiley Registry of Mass Spectral Data and the Mass Spectrometry Data Center data base produced by NIST–EPA, both use a Quality Index (QI) algorithm [9.18] (see Section 9.4.3).

9.4   Spectral Data Bases

Although there are over 1×10^7 known chemical compounds, most spectral data bases cover less than 100 000 compounds. The number of hard-copy compilations of spectral data is 20–30 times larger than the number of data bases [9.19]. Details of the main analytical data bases are given below. The software used to search the data bases is described in Section 9.4.5. Structure elucidation from spectra is covered in Section 11.6.7.

9.4.1   Nuclear Magnetic Resonance (NMR) Spectroscopy

Carbon-13 NMR Data Base on the Chemical Information System (CIS). This data base [9.20] was originally constructed by the Royal Dutch Chemical Society and currently contains only 11 700 spectra. The CIS [9.21], [9.22] is a collection of online data banks with the same chemical substructure search facility, the Structure and Nomenclature Search System (SANSS, see Section 9.6.3).

Carbon-13 NMR Data Base of the Fachinformationszentrum (FIZ). This data base has been available online on STN International since December 1987. It contained about 68 000 spectra of 60 000 compounds in 1988. The data base and retrieval software are related to those of the SPECINFO system (Section 9.4.5).

The sources of the data are journals, spectral catalogs, and unpublished spectra from BASF, ICI, and the Gesellschaft für Biotechnologische Forschung.

Data include chemical shifts, experimental conditions, literature references, molecular formulae, structures, coupling constants, and relaxation times. The database-specific software was developed at BASF with the following retrieval options:

1) Search of reference spectra by entering the lines of a measured spectrum
2) Search of reference spectra by entering a name fragment of a chemical compound
3) Search of spectra via full or partial molecular formula
4) Search of similar spectra
5) Search for structure fragments based on a measured spectrum
6) Search of compounds and their spectra with a defined structure or substructure
7) Estimation of chemical shifts for a defined structure

Bruker Spectroscopic Data Base is only available to users of Bruker spectrometers. It contains 19 000 carbon-13 NMR spectra and 900 proton NMR spectra.

Sadtler Laboratories Data Base. In 1989 this data base contained 28 000 carbon-13 NMR spectra which will soon be in full digital format, linked to a chemical substructure search system (see Section 9.4.5).

Collection of the National Chemical Laboratory for Industry (NCLI). This is part of the Japanese integrated online Spectral Data Base System [9.23]. It lists 6000 proton NMR spectra and 5700 carbon-13 NMR spectra. Software is provided for the user to look up or predict a spectrum, or match an unknown spectrum.

Other NMR data bases include the Varian data base. Tsukuba University (Japan) produces a collection of carbon-13 NMR spectra of polymers on CD-ROM.

9.4.2 Infrared (IR) Spectral Data Bases

Many large collections of IR spectra were built up when the spectrometers were prism and grating instruments. Fourier transform infrared (FT-IR) spectrometers, which generate digitized spectra, are now commonplace and new collections of data are being constructed.

Aldrich–Nicolet Digital FT-IR Data Base and the Sigma–Nicolet Biochemical Library. The Aldrich–Nicolet collection contained 10 600 compounds in 1987, and the Sigma–Nicolet library contained 10 400 compounds. The data bases are designed for use on personal computers. Software is supplied for matching peak intensities and locations from an unknown spectrum [9.24].

Sadtler Research Laboratories. This is the largest commercially available collection of IR spectra with about 60 000 spectra largely from prism and grating spectrometers. FT-IR spectra are being added.

Coblentz Society Spectra. The Coblentz Society has digitized 4400 of the spectra from the collection of 10 500 compiled and evaluated in collaboration with the Joint Committee on Atomic and Molecular Physical Data (JCAMP). The data base will be available for use with personal computers.

Clearinghouse for Digital IR Spectra. This project was initiated at the University of California–Riverside in October 1986 for the construction of a data base of digitized FT-IR spectra. It involves an automated algorithm for evaluating the spectra [9.25].

IR Data Committee of Japan (IRDC). This computer file has ca. 19 000 peak wavenumbers and intensities with search software requiring entry of wavenumbers and intensities in order of decreasing intensity. Spectra retrieved are listed in order of the probability of being a correct match. The possibility of fully digitizing the spectra has been discussed.

NCLI Japan. The Spectral Database System mentioned in Section 9.4.1 also includes 22 500 IR spectra [9.23]. All the spectra were determined at NCLI under carefully controlled conditions. The IR spectra were transferred to the data base directly, in digital form, from an FT-IR instrument. The data base is available online in Japan.

Other IR Spectral Data Bases. Some 3300 vapor phase spectra from the Environmental Protection Agency (EPA) laboratories are available through instrument manufacturers. Coded IR data on 145 000 compounds, compiled by the American Society for Testing and Materials (ASTM) are available from Chemir Laboratories and Sadtler Research Laboratories and online on the Canadian Scientific Numeric Data System.

9.4.3   Mass Spectral Data Bases

Mass spectral data bases are almost entirely devoted to electron impact (EI) spectra.

The National Institute of Standards and Technology, Environmental Protection Agency, Mass Spectrometry Data Center Data Base (NIST/EPA/MSDC Data Base). NIST used to be called the National Bureau of Standards (NBS) so the data base is still often referred to as NBS/EPA/MSDC.
The data base was originally constructed by S. HELLER of EPA and G. MILNE of the National Institutes of Health (NIH) [9.20],[9.26]–[9.29]. The Mass Spectrometry

Data Center of Nottingham later became involved. In November 1988 the data base consisted of 50 000 spectra, each one corresponding to a unique chemical compound.

Quality control of the chemical structural information [9.30] is achieved by cooperation with CAS in the allocation of CAS Registry Numbers of systematic names. The spectra are evaluated using a Quality Index (QI) algorithm [9.18] based on that developed by McLAFFERTY [9.31]. The NIST data base uses the following quality factors, each with a value between zero and one.

QF1 electron voltage
QF2 peaks at mass to charge ratio (m/z) above that corresponding to the molecular mass
QF3 illogical neutral losses
QF4 isotopic abundance accuracy
QF5 number of peaks in a spectrum
QF6 lower mass limit of the spectrum
QF7 sample purity
QF8 calibration date
QF9 similarity index of calibration mass spectrum

To obtain the Quality Index for the spectrum, all the quality factors are multiplied, then the product is multiplied by 1000. The Quality Index is calculated for each compound to be entered into the data base. If a new spectrum is discovered for an old compound, the new spectrum displaces the old one if it has a higher QI.

The data base is available online on CIS as MSSS (Section 9.4.5). It is distributed on tape without search software and in a PC version with software. The data base can be searched by identification number, CAS Registry Number, chemical name, molecular formula, molecular mass, and abundances of ten major peaks.

Wiley Registry of Mass Spectral Data. This collection is maintained by F. McLAF-FERTY at Cornell University. The data base, containing 123 704 spectra of 108 173 compounds, is available from J. Wiley and Sons on magnetic tape or CD-ROM. The magnetic tape version is distributed without search software, but software for matching unknown spectra is available free from Cornell University [9.32]–[9.36]. The data base is available online on CIS (Section 9.4.5).

In contrast to the NIST/EPA/MSDC data base, the Wiley data base has more spectra than compounds. A Quality Index algorithm is used, but is based on only seven Quality Factors (QF1–QF6 and a seventh called the source of the spectrum) [9.30].

The Eight Peak Index. This collection of partial spectra (65 000 eight-peak spectra for 52 332 compounds, indexed by molecular mass, chemical formula, and most abundant ions) is available on tape from the Royal Society of Chemistry.

NCLI (Japan). The Spectral Database System (see Sections 9.4.1 and 9.4.2) includes 10 000 mass spectra [9.23].

Japan Information Center for Science and Technology (JICST). The JICST runs an online mass spectral data base system searchable by name, formula, CAS Registry Number, and peaks. It is based on the NIST/EPA/MSDC data base, with 6000 additional spectra from the Mass Spectrometry Society of Japan.

MASS–LIB is a software package for automatic evaluation of low-resolution mass spectra. It was developed by the Max-Planck Institut für Kohlenforschung (FRG) and is sold by Chemical Concepts of Weinheim for use on VAX/VMS computer systems. It implements both the Wiley and NIST spectral data bases, allows input and maintenance of the user's own spectral libraries, and offers graphical representation of chemical structures.

9.4.4 Building Spectral Data Collections

Spectral data-base systems tend to specialize in just one particular technique. However, structure elucidation and other work need a multispectroscopy data bank. HELLER illustrates the lack of overlap of data banks with the diagram shown in Figure 3 [9.30].

Collections of data, such as those of Nicolet or NCLI, put together in a single laboratory by systematically determining spectra of compounds, are likely to be of high quality but data-base building is expensive and relatively slow. Collections put together through donations of spectra from laboratories can be built relatively quickly and inexpensively, but the data may be inaccurate. Collection of data from the literature (e.g., the Wiley registry) can quickly increase data-base size but many spectra are likely to be incomplete.

Most spectral data bases were started as a common need of a group of users. The most useful data bases have been created as a result of extensive organization and coordination and, most importantly, with adequate funding. The size and relevance

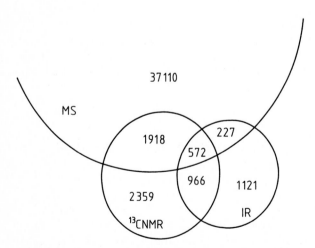

Figure 3. Lack of overlap in spectral data bases (reproduced with permission of Ellis Horwood [9.30])

of a data bank can be improved in a large industrial organization by integrating the sample-management and reference data-base systems, thus allowing internal data to be added to external data.

9.4.5   Spectral Search Systems

Until recently, most spectral data bases did not contain chemical structural data and did not have facilities for substructure searching. The CIS data bases with the SANSS software were an exception (see Sections 9.4.1 and 9.6.3). The software developed by BREMSER at BASF (see Section 9.4.1) offers a form of structure searching based on codes.

Sophisticated chemical structure handling systems that have been linked to spectral data systems include Molecular Design Limited's MACCS and the software used for CAS ONLINE data bases and FIZ's C13NMR on STN International [9.1], [9.7], [9.37], [9.38]. Two important microcomputer software packages are the Chemist's Personal Software Series (CPSS) from Molecular Design Limited [9.6], [9.7], [9.39], [9.40] and PSIDOM from Hampden Data Services [9.6], [9.7], [9.39], [9.41] (see Chap. 10).

In addition to searches for specific terms or compounds, and for spectra with certain peaks, there is a need for library or full-spectrum search, i.e., a best-match algorithm which finds those spectra that resemble most closely the spectrum of an unknown compound. Two well-known algorithms for whole spectrum searches of mass spectra are the Biemann KB, or forward search algorithm [9.42] and the McLafferty, Probability-Based Matching (PBM) algorithm [9.32], [9.33], [9.43].

When FT-IR digital spectra with accurate wavelength registration became available, full spectral search methods for IR spectra were also developed [9.24], [9.44]–[9.51].

Spectrum search of the carbon-13 NMR data base on CIS uses an algorithm developed by CLERC and coworkers [9.52]. The SPECINFO and CSEARCH systems also offer search software.

The following examples illustrate systems where chemical substructure searching is linked to a spectral data system. Systems for handling chemical structures are described in Chapter 10.

Spectral Data Bases on CIS. The CIS offers four spectroscopy data banks:

1) CNMR Carbon-13 NMR Spectral Search System
2) IRSS Infrared Search System
3) NMRLIT Literature Search System
4) WMSSS Wiley Mass Spectral Search System incorporating the MSSS spectral data file from NIST

Chemical structures are searched using the Structure and Nomenclature Search System, SANSS [9.53]. Further information on CIS is given in Section 9.6.3.

CNMR permits search by chemical shift, and analysis and display are available.

IRSS permits retrieval of known spectra and analysis of unknowns. The data base includes approximately 3000 spectra taken from the EPA collection plus contributions from the Boris Kidric Institute in Yugoslavia. A graphics package allows display of spectra on graphics terminals. WMSSS contains both the NIST/EPA/MSDC and the Wiley mass spectral data bases. The spectra can be searched on the basis of peak and intensity requirements, as well as by Biemann and probability-based matching (PBM) techniques. Searches by molecular mass, molecular formula, and partial formula are also permitted.

NMRLIT permits searching the index to Nuclear Magnetic Resonance Literature Abstracts and Index (Preston Publications, Inc.). Currently there are over 43 000 references covering abstracts published from 1964 through December 1984. Searches may be on subject, nucleus, author, or general reference.

SPECINFO is a structure-related, multidimensional spectroscopic data base system for VAX computers that was written by Bremser and coworkers at BASF [9.54]–[9.57]. Since January 1990 it has been marketed by Chemical Concepts of Weinheim (FRG). Not only is SPECINFO multidimensional (NMR, IR, and mass spectra), but it also incorporates chemical structure handling software. The carbon-13 NMR data base and similar software are available online on STN International. The data available in SPECINFO are shown in Table 8.

The in-house system uses color graphics displays. Full connection tables are stored for each compound but structure searching is done by HOSE (Hierarchically Ordered Spherical Description of Environment) codes or HORD (Hierarchically Ordered Ring Description) ring codes stored in inverted files. Structure queries are entered by naming various fragments and giving numbers for ring positions. The input is textual but the structure is displayed graphically.

The data bases can be searched for a variety of single parameters (e.g., molecular formulae, NMR coupling constants, mass spectral masses), CAS ring index codes, and compound names. Spectral data can be displayed both graphically and as a listing.

Table 8. Data available on SPECINFO*

Type of spectra	Number of spectra and structures
C-13 NMR	100 000
P-31 NMR	2 000
O-17 NMR	800
N-15 NMR	1 000
H-1 NMR	20 000
Mass spectra	3 000
Mass spectra (NIST)	40 000
IR	16 500

* Wiley mass spectra are not incorporated.

Input of spectral data is somewhat laborious at present. Registration of additional spectra has recently been made available. Automatic search of incoming data is now a facility.

The only NMR search available at present is a sequential line search but SAHO (Spectral Appearance in Hierarchical Order) and similar search features (available on STN International) will eventually be added. SPECINFO incorporates useful NMR spectral prediction and coupling constant prediction software. There are plans to link SPECINFO with MASS-LIB.

MASS-LIB. MASS-LIB is a set of programs developed by HENNEBERG of the Max Planck Institut für Kohlenforschung (FRG) for the evaluation of low-resolution mass spectra [9.58]. It was sold by MSP Friedli and Co. (Switzerland) for use on VAX/VMS computer systems. From 1st January 1991, Chemical Concepts has the right to sell MASS-LIB. It implements both Wiley and NIST mass spectral data bases and allows users to maintain their own libraries. Structures can be displayed but not searched. Compared with SPECINFO, the strength of MASS-LIB lies in its SISCOM (Search for Identical and Similar Components) library search and in its larger data base. Its storage algorithm allows almost the whole spectrum to be used for comparison without an unacceptable overhead in terms of speed of search.

CSEARCH. CSEARCH has been developed over 11 years by ROBIEN and coworkers at the University of Vienna [9.59] and is now being marketed by Sadtler. The combined carbon-13 NMR academic and commercial data base available through Sadtler contains 52 000 chemical entries with spectra covering 11 nuclei. The system allows users to read a spectrometer-generated peak list directly; to search spectra for similar or identical shift patterns; to search by either full structure or substructure; to combine chemical shift and structure searches; and to estimate spectra from structure proposals.

CSEARCH currently runs on the Silicon Graphics family of workstations and is currently being ported to X-Windows so that it will be able to run in other work-station environments.

There is an interface to Molecular Design Limited's program ChamBase (see Section 10.6.3). PC-based data addition software is also available.

Fluorine-19 NMR Data Base. Fraser Williams (Scientific Systems) Ltd. has written special software to go with its fluorine-19 NMR data base. When structures and spectral data are displayed together, the relevant atoms are highlighted by dark "blobs". The data base plus PC-SABRE search software are available for use on an IBM PC [9.6].

Sadtler PC Spectral Search Libraries. Sadtler's PC SEARCH software operates on IBM PC/AT and PS/2 computers. Full spectrum search (IR and carbon-13 NMR) is possible and the hits are listed in order of closeness of matching. Two IR spectra can be displayed in different colors on one screen and the relevant chemical structures shown. A difference spectrum can be obtained.

The software operates under the application environment Microsoft Windows. The chemical structures for the 59 000 compounds for which IR spectra are available have been registered in a PsiBase data base [9.6], [9.7], [9.39], [9.41]. Substructure searching is therefore possible and a hit list from a PsiBase search can be taken into PC SEARCH for spectral search and display.

There will also be a 28 000 compound carbon-13 NMR data base and possibly a UV–visible spectral data base in the same system.

9.5   Crystallographic Data Bases

More than 100 000 full three-dimensional (3-D) crystal structure analyses have been reported, of which over 50% have been published since 1980 [9.60]. Because of the complexity of the atomic coordinate data, they cannot all be recorded in the primary literature. Molecular graphics data and theoretical chemistry data (e.g., molecular orbitals) are particularly well suited to data bank storage. Although printed handbooks of crystallographic data are available, online data banks have become the main source of three-dimensional data and provide the information in a comprehensive, versatile, and readily accessible form. Sophisticated software enables structures to be retrieved, manipulated, and compared.

The Cambridge Structural Database (CSD). The Cambridge Crystallographic Data Center (CCDC) was set up in 1965 to compile a computerized data base on 3-D molecular structures. The CSD is the world's largest data base of evaluated experimental results, it contains over 90 000 compounds with the addition of about 8500 per year (1990) [9.1], [9.60]–[9.62].

All information is thoroughly checked and evaluated before being incorporated. Numeric data are also subjected to computerized internal consistency checking (see Section 9.3). The percentage of entries still in error in the CSD is 1%, compared with ca. 15% in the literature [9.63].

Bibliographic (1-D), connectivity (2-D) and 3-D structural data are combined in one data base named ACER [9.64]. The search software, QUEST, operates on ACER and outputs four subfiles, REFC (reference numbers), FCON, FBIB and FDAT as well as a subset of the data base in the CSD format ACER. FCON and FBIB contain connectivity and bibliographic data respectively. FDAT contains the 3D data which can be searched using a programm called GSTAT and displayed by a program PLUTO. Version 4 of the CSD system (see Section 10.5.4), released in 1989, uses high quality graphics for query input and for the display of "hits", and the data base is augmented by two-dimensional structural diagrams for 90% of the compounds. The search output has been improved by highlighting the substructural search fragment.

The Protein Data Bank. The Protein Data Bank, produced by the Brookhaven National Laboratory, is supported by the U.S. National Science Foundation and

NIH. It was established in 1971 as a computerized archive for three-dimensional structures of biological macromolecules [9.60], [9.65].

Over 1200 biological macromolecules have been crystallized, of which about 400 structures have been determined. In 1987, PDB held about 350 coordinate entries, with bibliographic references for most of the remaining structures. The format adopted by the PDB has been accepted as a standard for the interchange of atomic coordinates [9.66].

NBS Crystal Data. This data base is produced with the support of the NIST Office of Standard Reference Data. There are about 125 000 entries covering crystallographic, chemical, and physical data, with over 7000 annual additions. Each data entry is critically evaluated at NIST before being incorporated. The data base is available online through CRYSTDAT by the cooperation of the Canadian Institute for Scientific and Technical Information (CISTI), the Canadian Scientific Numeric Database Service (CAN/SND), and the NBS Crystal Data Center (USA). Although full structure searching is not possible, lattice-formula search techniques [9.67] enable data for unknown compounds to be matched with entries in the data base and the compound can be identified.

CAN/SND also provide access to the Cambridge Crystal Structure Database, the Protein Data Bank, the Inorganic Crystal Structure Database, and the NRCC Metals Crystallographic Data File also described in this Section.

Other Crystallographic Data Bases [9.60]. The *Inorganic Crystal Structure Database (ICSD)* is produced by the Gmelin Institute in cooperation with the University of Bonn [9.68]. It contains 29 000 records (1990) of inorganic crystal structures and can be searched via STN International.

The *NRCC Metals Crystallographic Data File (CRYSTMET)* contains critically evaluated crystallographic and bibliographic data for metallic phases, determined by diffraction methods. The file is maintained by the Canada Institute for Scientific and Technical Information (CISTI) through the Scientific Numeric Database office (CAN/SND). The data base contains about 11 000 entries for compounds where the composition is clearly defined and the space group and unit cell have been determined.

The creation of a *Biological Macromolecule Crystallization Data Base* is supported by NIST. To be included, the data must have been published and the macromolecules must have at least one crystal form for which the unit cell and space group have been established by X-ray diffraction analysis.

The *Powder Diffraction File* is distributed and maintained by the Joint Committee on Powder Diffraction Standards (JCPDS)-International Center for Diffraction Data, Pennsylvania, and is a collection of single-phase X-ray powder diffraction patterns. It contains over 44 000 patterns, with the addition of about 2000 per year.

The *Crystal Data Identification File (CDIF)* is owned by the International Center for Diffraction Data (USA) and contains crystal class and unit-cell information for about 60 000 structures. All cell data are reduced to a standard form, and software allows searching on the basis of cell dimensions.

9.5.1   Molecular Sequence Data Banks

A chapter on chemical data banks would be incomplete without reference to the growing number of protein sequence data banks. Over 10 000 protein sequences have been determined so far [9.69]. To cope with the task of assembling evaluated data, CODATA recommended in 1984 that three protein data bases should cooperate to form the Protein Identification Resource (PIR) International Association of Protein Sequence Data Collection Centers [9.70]. The three data bases are the PIR at the National Biomedical Research Foundation (PIR–NBRF) in the United States, the Martinsried Institute for Protein Sequences (MIPS) in the Federal Republic of Germany, and the Japanese International Protein Information Databank (JIPID).

The National Library of Medicine (NLM) is also cooperating internationally to collect molecular sequence data; 13 data banks register molecular sequences deposited with them by researchers [9.71].

Chemical Abstracts Service are setting up two data bases (CASSEQ–Nucleic Acid and CASSEQ–Protein) to hold all the sequences in their abstracted references [9.72].

9.6   Chemical and Physical Property Data Bases

Until the advent of computer data banks, the chemical and physical properties of compounds were to be found in one of the encyclopedias or handbooks described in Chapter 3. Many of these are now available online, the most comprehensive being the Beilstein handbook, available as Beilstein Online (Section 9.6.1) [9.11], [9.12], [9.73], [9.74]. Many other data banks containing chemical and physical properties of compounds are offered online but most are independent of one another and searches cannot be linked. The Chemical Information System (CIS), however, is an integrated collection of over 30 data banks [9.75], [9.76], linked by a central index file, the SANSS file [9.53] (Section 9.6.3).

9.6.1   Beilstein

The Beilstein Handbook is the world's largest compilation of evaluated data on organic chemistry (see Section 3.3.2). Beilstein Online, first released in 1988, provides more up-to-date information than the printed version, and also enables searches to be carried out in English. It is available on STN International, DIALOG, and ORBIT (Maxwell Online) and is expected to be available on Data Star.

The data base contains two types of records [9.12]:

1) Organic substances cited in the Beilstein Handbook, Basic Series and Supplementary Series I–IV (1830–1960). These records constitute the *Full File*.
2) Organic substances from the Beilstein collection of excerpts from the primary literature (1960–1980) that have not yet been critically reviewed and are being published as Supplementary Series V. These records constitute the *Short File*.

Information in the Full File has been checked for errors and redundancies, but the Short File is created from data abstracted directly from the primary literature without checking. Data in the Short File will be added to the Full File when it has been checked.

For each compound, information is stored in the structural and factual files. Structure searching is covered in Chapter 10. The data in the factual file can be divided into seven parts:

1) Identifiers: molecular formulae and registry numbers for identification and search of the compounds
2) Structure-related data: information on the purity of the compound and alternative structure representations
3) Preparative data: yield, solvents, temperature, pressure, etc.
4) Physical properties: structure and energy parameters; physical properties of the pure compound; physical properties of multicomponent systems
5) Chemical behavior: reactions with other chemicals
6) Physiological behavior and applications: toxicity, biological functions, ecological data
7) Characterization of derivatives and salts

There are over 500 fields, of which more than 300 are searchable. The user can search for structures and substructures using graphical input or string search; for numerical terms using Boolean logic; for keywords; or for other key fields such as molecular formula, CAS Registry Number, or Beilstein Registry Number.

9.6.2 Other Handbooks and Encyclopedias Available Online

Several of the reference works described in Chapter 3 are available online [9.77] and cover chemical and physical properties. The HEILBRON data base, available on DIALOG since 1986, corresponds to the Dictionary of Organic Compounds, Fifth Edition (DOC5) and six related dictionaries (Section 3.3.3) [9.9]. HEILBRON contains physical and chemical data plus bibliographic references to 250 000 substances.

Other reference works available as online data banks include:

1) The Kirk–Othmer Encyclopedia of Chemical Technology, available via DIALOG and BRS (Maxwell Online).
2) The Merck Index, available on DIALOG, BRS, and CIS.
3) The Formula Index (GFI) and the Complete Catalog from the Gmelin Handbook of Inorganic and Organometallic Chemistry, accessible via STN International. The Gmelin Institute is also preparing a Factual Data Bank.
4) HODOC, produced by CRC Press and containing information from the second edition of the Chemical Rubber Company's Handbook of Data on Organic Compounds (HODOC II), available online via STN International.

9.6.3 Data Banks on the CIS System

The Chemical Information System (CIS) is a collection of about 30 data banks developed originally by the U.S. government agencies NIH and EPA, and now marketed by CIS Inc., a subsidiary of Fein–Marquart Associates. The independent data banks are linked through a central file, the Structure and Nomenclature Search System (SANSS). Chemical structure searching is discussed in Chapter 10.

Over 350 000 chemical substances are searchable by full structure, substructure, or by name or partial name. Structure query entry can be facilitated with the front-end package, SuperStructure [9.6]. Most of the data banks can be searched using a common command language Text Data Retrieval System (TDRS), which also allows transfer of data from one data bank to another. The compounds are identified and cross-referenced by their CAS Registry Numbers.

The information on the data banks can be classified into:
1) Spectral data banks (Section 9.4.5)
2) Chemical and physical property data banks (Table 9)
3) Toxicology, environmental, and hazard data (Section 9.7)
4) Pharmaceutical data banks
5) Analysis and modeling data banks.

Table 9. CIS data banks of physical and chemical properties

Name of data bank	Producer	Number of substances covered	Type of data available
CESARS (Chemical Evaluation Search and Retrieval System)	Office of Materials Control, State of Michigan	194	physical and chemical properties, carcinogenicity and toxicity, environmental fate, hazard and handling information
CHRIS (Chemical Hazard Response Information System)	U.S. Coastguard	1 156	labeling, physical and chemical properties, health hazards, fire hazards, chemical reactivity, water pollution
ENVIROFATE (Environmental Fate)	EPA Office of Toxic Substances, Syracuse Research Corporation	450	physical and chemical properties, environmental transformation rates
ISHOW (Information System for Hazardous Organics in Water)	EPA Environmental Research Laboratory, University of Minnesota, EPA Office of Toxic Substances	5 400	melting points, boiling points, partition coefficients, acid dissociation constants, water solubility, vapor pressures
OHM/TADS (Oil and Hazardous Materials/Technical Assistance Data System)	EPA	1 400	physical, chemical, biological, toxicological, and commercial data
SOLUB (aqueous solubility data base)	University of Arizona College of Pharmacy	2 600	evaluated data on aqueous solubility
THERMO (thermodynamics data base)	NIST	15 000*	evaluated thermodynamic data

* Entries.

The pharmaceutical data bases are: Drug Information Fulltext (DIF), which contains the complete contents of two reference volumes published by the American Society of Hospital Pharmacists; The Merck Index Online (see Section 9.6.2); Physicians' Desk Reference Online, which contains the full text for prescription, over-the-counter, and ophthalmological drugs.

The analysis and modeling data banks are: ARTHUR, which is used in the formulation and evaluation of models for incompletely understood data sets; CHEMLAB (Chemical Modeling Laboratory), which provides capabilities for three-dimensional conformational analysis, molecular orbital calculations, and estimation of many chemical properties; MLAB/CLAB (Mathematical Modeling System including Cluster Analysis Laboratory), which is an interactive system for mathematical modeling, cluster analysis, and pattern-recognition of numeric data.

9.6.4   Thermodynamics and Thermophysical Property Data Banks

Of all the chemical properties, thermodynamics is the most important for many scientists [9.78]. It is particularly important that thermodynamics data bases contain accurate, validated information, as the searcher is likely to calculate other properties from retrieved data.

STN International Data Bases. There are three thermodynamic data bases on STN International: DIPPR, JANAF, and NBSTHERMO. STN International provides access to a number of numeric data bases [9.77], [9.79]. Several of these have already been described: the carbon-13 NMR data base (Section 9.4.1), ICSD (Section 9.5), Beilstein Online (Section 9.6.1) and HODOC (Section 9.6.2). Numeriguide, produced by the American Chemical Society, is a compilation of the lists of properties in each of the STN numeric files.

DIPPR contains physical property data, textual information, and CAS Registry Numbers for over 1000 commercially important chemical compounds. Data are compiled and evaluated by the Design Institute for Physical Property Data (DIPPR) of the American Institute of Chemical Engineers. DIPPR is also available on Numerica.

JANAF contains the Joint Army, Navy, and Air Force Thermochemical Tables and is produced by the U.S. Department of Commerce, NIST. There are 1100 records of critically evaluated chemical thermodynamic properties of inorganic substances.

NBSTHERMO is also produced by NIST and corresponds to the data in the NBS Tables of Chemical Thermodynamic Properties. It contains over 8200 records of critically evaluated chemical thermodynamic properties of inorganic and organic substances.

Numerica Data Bases. Technical Database Services, Inc. of New York also provide thermodynamic data bases through their Numerica service.

Data Files from the Thermodynamics Research Center (TRC). The TRC at Texas A&M University has created two numerical data bases of thermodynamic properties [9.80]: the TRC Selected Data File, which contains selected values of properties published in the TRC Thermodynamic Tables, and the TRC Source Data File, which contains properties of pure organic compounds that have been published in the open literature.

The TRC Thermophysical Property Datafile 1: Vapor Pressure is available online through Numerica. It contains about 18 000 records of experimental data for 3500 chemicals. Vapor pressure and boiling points for over 5500 chemicals can be calculated interactively using critically evaluated Antoine coefficients.

The Physical Property Data Service (PPDS) is an integrated system which provides thermodynamic and transport property data for pure components and mixtures. It is maintained at the National Engineering Laboratory and the Institute of Chemical Engineers (UK) and is available online through Numerica. Supplementary programs allow the users to incorporate their own data.

DECHEMA Thermophysical Property Data Bank (DETHERM). The Deutsche Gesellschaft für Chemisches Apparatewesen, Chemische Technik and Biotechnologie e.V. (DECHEMA) produces a number of reference works, including the DECHEMA Chemistry Data Series [9.81].

The DETHERM data bank is produced in conjunction with FIZ CHEMIE and is available online via Numerica. It includes three systems:

1) DETHERM-SDR (Data Retrieval System),
2) DETHERM-SDC (Data Calculation System), and
3) DETHERM-SDS (Data Synthesis System). DETHERM-SDR is a compilation of physical property data on 3000 components abstracted from the literature. DETHERM-SDC is used to compute properties and phase equilibria from the stored data. DETHERM-SDS can be used to create new pure-component data for subsequent use with the SDC programs.

Other Thermodynamic and Thermophysical Property Data Banks. The THERMO data base on CIS is included in Table 9. The data base F*A*C*T (Facility for the Analysis of Chemical Thermodynamics) uses some of the same data sources as THERMO, although it only contains data on 2400 substances. The user can calculate thermodynamic equilibria or phase diagram boundaries.

Thermodata (France) produces a bibliographic data base Thermodoc and three thermodynamic data banks: Thermocomp, Thermalloy, and Thermosalt.

Thermodynamic properties of organic substances can be calculated using data from the data base EPIC (Estimate of Properties for Industrial Chemistry) available online through CIGL Inc. in Belgium.

Online data bases of vapor–liquid equilibrium data [9.76] are CHEMTRAN and the Dortmund VLE Data Bank (ChemShare Corp.), and the Physical Property Data Service (PPDS) produced by the Institute of Chemical Engineers (UK) and available through Numerica.

9.6.5   Other Chemical and Physical Property Data Bases

The Pesticides Properties Database is being developed by the USDA Agricultural Research Service (ARS) to provide essential data on the chemical and physical properties of pesticides [9.15]. All the data obtained from the literature are critically evaluated. The data base will help in regulatory assessment of pesticides and in developing models for determining the contamination of drinking water supplies by pesticides entering the soil.

Codura Publications Inc. produce Poly-Probe, a data base of physical property data on plastics.

9.7   Toxicology, Hazard, and Environmental Data Banks

Toxicology data are very important [9.82], but there are more problems associated with their use and creation than with most other types of data bank. It is difficult to use a toxicology data bank without a good knowledge of the subject and some experience in the use of information systems. There are problems with evaluation and quality control of data. Checks for internal consistency are difficult and repeating the measurements is not normally feasible.

CIS Toxicology Data Banks. The Registry of Toxic Effects of Chemical Substances (RTECS) is compiled by NIOSH and is the largest chemical toxicology data bank, containing information on 90 000 substances. Data cover acute and chronic toxicity measurements; primary skin and eye irritant data; carcinogen, mutagen, and tumorigen data. Negative findings are not recorded and the data are not evaluated. Thus, RTECS cannot necessarily be used as a substitute for the original source. However, the linking of chemical substructure searching with data searching allows correlation of structure and toxic effects. The data bank is available on CIS, and as part of the CHEM-BANK CD-ROM from SilverPlatter.

CIS contains several other data banks that include toxicological, environmental, and hazard information [9.75]; CESARS, CHRIS, ENVIROFATE, ISHOW and OHM/TADS are listed in Table 9 (p. 88). The six remaining toxicological data banks are shown in Table 10. In addition, CIS provide a bibliographic data base on GastroIntestinal Absorption (GIABS) and an index to unpublished health and safety studies submitted under the Toxic Substances Control Act (TSCATS).

Chemical Hazards Data Banks. Most of the CIS data banks relating to hazards have already been covered: RTECS, OHM/TADS, CHRIS and CESARS (Tables 9 and 10). CIS also provide the HAZINF data bank (Hazardous Chemicals Informa-

Table 10. CIS toxicological data banks

Name of data bank	Producer	Number of compounds covered	Type of data available
AQUIRE (Aquatic Information Retrieval)	EPA Environmental Research Lab, EPA Office of Toxic Substances	4 000	acute, chronic, bioaccumulative, sublethal effects data from experiments performed on freshwater and saltwater organisms
CCRIS (Chemical Carcinogenesis Research Information System)	National Cancer Institute	1 000	assay results and test conditions for carcinogenicity, mutagenicity, tumor promotion, and cocarcinogenicity
CTCP (Clinical Toxicology of Commercial Products)	5th edition of publication by Gleason, Hodge, Gosselin and Smith	20 000*	toxicity data for about 20 000 common commercial products, including symptom and treatment data
DERMAL (Dermal Absorption)	EPA Office of Toxic Substances	655	data on toxic effects, absorption, distribution, metabolism and excretion related to dermal application of chemical substances
GENETOX (Genetic Toxicity)	National Institute of Environmental Health, Oak Ridge National Laboratory, Environmental National Laboratory	2 600	mutagenicity information tested against 38 biological systems
PHYTOTOX (Plant Toxicity)	University of Oklahoma, Dept. of Botany and Microbiology; EPA	1 000	biological effects of application of chemicals to plants
RTECS (Registry of Toxic Effects of Chemical Substances)	NIOSH	90 000	acute and chronic toxicity data

* Commercial product.

tion and Disposal Guide). This contains hazard ratings, chemical and physical properties, hazardous reactions, fire hazards, physiological properties, health hazards, and waste and spillage disposal instructions for about 220 chemicals or classes of chemicals. The data base was created by the Department of Chemistry, University of Alberta.

A comparison of the coverage of chemical hazard data banks is given in [9.83]. The Hazardous Substances Data Bank (HSDB), produced by the National Library of Medicine, Toxicology Information Program, was found to have the widest subject coverage and is a good source for human toxicity information. Although it only contains approximately 6500 compounds, much of the data is evaluated [9.82].

The National Chemical Emergency Center at Harwell (UK) produces the microcomputer system CHEMDATA, which provides information for identifying hazards and procedures for dealing with chemical products in emergencies. Safety measures for 18 000 substances are available. CHEMDATA is one of the seven sources of information on the Dangerous Goods CD-ROM from Springer Verlag. Also available on the CD-ROM are:

1) Hommel Handbook of Dangerous Goods
2) The Operation Files for chemical incidents of the Swiss Fire Brigades
3) The Fluka Catalog

4) The Swiss Toxic Substances List, containing 76 000 substances
5) SUVA, the Swiss Accident Insurance Institution, which contains the safety codes of liquids and gases
6) VCI, the handbook of firms in the chemical industry (Econ-Verlag, FRG)

The SilverPlatter CD-ROM, CHEM-BANK, contains information on hazardous chemicals (the RTECS, CHRIS, and OHM/TADS data banks). The data bank TOSCA (Toxic Substances Control Act) Initial Inventory from the EPA is to be added.

The Environmental Chemicals Data and Information Network (ECDIN). The ECD-IN data bank is produced at the Joint Research Center of the Commission of the European Communities at Ispra in Italy [9.84]. It contains data related to actual or potential chemical pollutants. The broad spectrum of parameters and properties supports the evaluation of the risks connected with the use of the substances. Links are established between various types of data, enabling the data bank to be used for correlation analyses. Public access is via the DC Host Center in Copenhagen and the data are managed under the ADABAS system. Chemical structure handling uses a modified form of the CROSSBOW system (Section 10.1.3), but chemical names and synonyms are the most common way of retrieving structures.

One of the ECDIN files is the European Inventory of Existing Chemical Substances (EINECS) which gives the chemical substances that were on the EEC market between January 1971 and September 1981. All these substances have been registered with CAS and their Registry Numbers are available on ECDIN. Substances not in the EINECS inventory must be notified before they can be marketed in the EEC.

Data Banks Produced as a Result of Legislation. Legislation has been the impetus behind the production of a number of data banks (e.g., EINECS). The CIS data bank TSCAPP was compiled as a result of the Toxic Substances Control Act which required the establishment of an inventory of chemicals in U.S. commerce from 1975 through 1977. The TSCAPP file contains plant and production data for over 55 000 chemicals on the Toxic Substances Control Act Inventory. The TSCATS file on CIS (Toxic Substances Control Act Test Submissions) contains about 15 000 records to over 1200 compounds.

Four other CIS files were created as a result of legislation:

1) *Federal Register Search System (FRSS)* provides a cross-reference to all citations to a chemical or class of chemicals cited in the Federal Register from 1 January 1977 to November 1983, but is no longer being updated.
2) *Suspect Chemicals (SUSPECT)* contains regulatory and advisory data for ca. 4000 potentially hazardous chemicals, including explosive limits. The information is taken from U.S. government regulations, National Fire Protection Hazard Ratings (NFPA), and Threshold Limit Values and Biological Exposure Indices (ACGIH).
3) *Chemical Activity Status Report (CASR)* lists chemicals that are studied by the EPA in the course of regulatory or scientific research activities.
4) *Industry File Index System (IFIS)* provides summaries of EPA regulations on particular industries and chemical substances.

Material Safety Data Sheets (MSDSs) are required to accompany the transfer of chemicals from one company to another. CIS provides access to more than 1600 MSDSs from J.T. Baker Co. and more than 1400 MSDSs from Mallinckrodt Inc. The MSDSs are formatted according to the OSHA guidelines, and contain information on safe handling, storage, and disposal.

Sigma–Aldrich Chemical Co. produce a CD-ROM of 30 000 MSDSs for about 15 000 compounds [9.85].

9.8   Special Applications

The ChemQuest data base on ORBIT (Maxwell Online) is created from the catalogs of 54 chemical suppliers. This fine chemicals directory can be searched by chemical name, trade name, molecular formula, CAS Registry Number, or by structure or substructure. Compounds can be ordered online.

A fine chemicals data base is available as the CD-ROM ChemLink data base from Chemron Inc. [9.85] and online on DIALOG. Catalogs from only 31 suppliers are included, but it is cheaper to use than ChemQuest.

9.9   The Future of Data Banks

In the field of chemistry, future progress should be directed towards the improvement of data quality and better integration of existing systems. Several barriers to building chemical data bases have been identified [9.19]; scientists have little experience with data-base management systems; most data bases are built for individual use and are hard to adapt to more general use; users have not been involved enough in the design of data bases; very few ways for accessing data bases are available to the chemical community at large; funding for data-base building rarely is available. Another problem is the lack of skill and interest to generate and use these data bases [9.20].

Most of the data banks described in this chapter have developed independently without standardization. Work on standards is being carried out by the National Standard Reference Data System which was established in the United States in 1963 to coordinate production and dissemination of critically evaluated data in the physical sciences. IUPAC is concerned with collecting and compiling numeric data sources; the Committee on Data for Science and Technology (CODATA) has been working on standards since 1966 [9.86]. World Data Centers (WDC) have existed since the 1950s [9.87]. More international cooperation is needed in the development of online data banks if the vast amount of data being produced is to be made available in a usable form. The use of CD-ROM as a storage medium has great potential in the dissemination of data banks.

10   Chemical Structure Handling

10.1   Chemical Structure Representation [10.1]–[10.4]

10.1.1   Systematic Nomenclature

The chemist's preferred method of communicating structural information is the two-dimensional (2-D) chemical structure.

In all chemical information systems there is a need to reduce the structure to a "name" that can be stored and manipulated by a computer. Systematic nomenclature (see Section 2.2.4) in its usual sense is not useful here for many reasons. Systematic names are often long and complicated. When a new area of chemistry is opened up the appropriate nomenclature is not available for months or even years. Systematic rules are often inconsistent or ill-defined. In IUPAC nomenclature there is inconsistency in the retention of trivial names and there may be more than one acceptable name for a single compound. In a useful computerized system any one compound must have only one name and any name must be convertible to only one structure. CAS has systematized nomenclature sufficiently to give a unique name for most compounds but has had to make its own rules. Trivial names are still in common usage for some applications.

Thus, while systematic nomenclature does have advantages, for example, in printed indexes, the chemical information specialist has turned to other methods for the internal representation of structures in the computer.

10.1.2   Fragmentation Codes

A fragmentation code is a collection of predetermined small substructures (e.g. six-membered ring, carbonyl group) each represented by a number, or sometimes a combination of letters and numbers. In the early days each fragment could be represented by a hole in a punch card. Thus a hydroxy group might be represented by punching a hole in the third row of the seventieth column of an 80-column punch card, represented on paper as 70/3. A punch card with appropriate holes was generated for each compound and the structure drawn by hand on the card. Mechanical card-sorting equipment (or even simple rods) were then used to find all the cards with a hole 70/3 (or any other required holes) to perform a search for hydroxy compounds.

Nowadays the codes are automatically derived from a form of computer representation and searched by a computer program.

There are two types of fragmentation code: fixed and open-ended. A *fixed fragmentation code* consists of a fixed number of fragments, and is not easily modified if new features need to be included. If a new fragment is invented, all previously processed structures will need recoding.

The fragments of an *open-ended code* are constructed from a structure representation by a computer algorithm according to a particular set of rules. Consequently, an open-ended code can be more responsive than a fixed code to the changing characteristics of the compound collection. Also, more precise retrieval from the file may be possible because the code offers a much greater variety of search terms.

Derwent Publications manually encodes structures into two fixed fragmentation codes: the Ring Code [10.5] and the New Chemical Code [10.6] for use with Derwent's RingDoc and World Patents Index Services, respectively.

An important and sophisticated, open-ended fragmentation code is GREMAS (Generic REtrieval by MAgnetic tape Search) [10.7]–[10.9] developed by the Internationale Dokumentationsgesellschaft für Chemie (IDC), a consortium of seven German companies (see Section 20.4). This code involves fragment terms which are constructed according to a number of broad structural classifications, each of which can be subdivided using a three-level hierarchy. These classifications can be used to characterize fragment occurrences at varying levels of specificity and the way in which terms can be combined, thus reducing the ambiguity of the representation and increasing the efficiency of searching.

Fragmentation codes are inevitably an ambiguous method of structure representation. They describe the various parts of a molecule but not how those parts are linked together (i. e., the full topology). Now that other methods of structure representation are available allowing the description of the full topology, fragmentation codes are used mainly for the initial screening stage in substructure searching (see Section 10.3).

10.1.3   Linear Notations

Linear notations are a compact way of representing a chemical structure by strings of alphanumeric characters. Two notations, Wiswesser Line Notation (WLN) [10.10]–[10.12] and Dyson/IUPAC [10.13] are compared in Figure 4.

WLN was the best known of the notations in the chemical industry for many years and was used by ISI for its Index Chemicus Registry System [10.14]. With a few minor exceptions, WLN is an unambiguous representation: it fully represents the constitution and connectivity of a molecular structure. Moreover, if the encoding rules are followed precisely, the notations are unique (only one linear string is

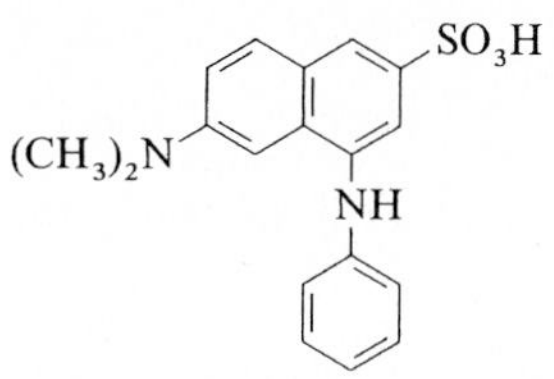

Figure 4. Comparison of Wiswesser Line Notation and Dyson/IUPAC Notations

possible for any compound), which makes WLN appropriate for registration (see Section 10.2) and the production of printed listings.

The ICI CROSSBOW system (Computerized Retrieval of Organic Structures Based on Wiswesser) used not only WLN but also fragment codes and connection tables (see Section 10.1.4) based on WLN [10.15]–[10.18]. CROSSBOW was widely used worldwide until the early 1980s.

WLN has some of the disadvantages of systematic nomenclature; it is not suitable for handling tautomerism, mesomerism, and stereochemistry. One of its principle advantages, compactness, has disappeared with the advent of low-cost, high-density storage. Its use has therefore given way to the use of connection tables (see Section 10.1.4). Many companies have converted their WLN files to connection table files using programs such as DARING [10.19].

A notation still in use by a select group is SMILES (Simplified Molecular Input Line Entry System) [10.20]. Its use with software supplied by Daylight Chemical Information Systems is described in Section 10.4.2.

10.1.4 Connection Tables [10.3], [10.4], [10.21]

A connection table contains a list of all of the non-hydrogen atoms within a structure, together with bond information that describes the exact manner in which the individual atoms are linked together. Hydrogen atoms are not included since their presence or absence can be deduced from the bond orders and atomic types.

Many different types of connection table are used. In a *redundant* connection table each bond appears twice (Fig. 5); in a *nonredundant* connection table each linkage is listed only once (Fig. 6). In *bond-implicit* connection tables (e.g., the CROSSBOW system), the bonds are not defined explicitly but are included in the "nodes" which characterize each atom. Figures 5 and 6 both show *bond-explicit* connection tables. They take up more storage space than bond-implicit tables but the information can be interpreted more easily by a computer program.

Connection tables are unambiguous since they give a complete description of the topology of the compound which they represent. However, they are generally not unique since there are very many different ways of numbering the atoms. There are algorithms to produce a unique connection table from a randomly numbered version. This process is usually known as "canonicalization", although the correct

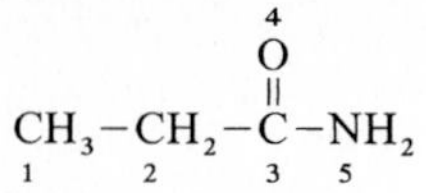

$$\overset{\overset{\overset{\displaystyle 4}{\displaystyle O}}{\displaystyle \|}}{CH_3-CH_2-\underset{3}{C}-NH_2}$$
$$\underset{1}{}\ \underset{2}{}\ \ \ \underset{5}{}$$

Atom number	Atomic symbol	Bond connection	Attached atom number	Bond connection	Attached atom	Bond connection	Attached atom
1	C	1	2				
2	C	1	1	1	3		
3	C	1	2	2	4	1	5
4	O	2	3				
5	N	1	3				

Figure 5. A redundant connection table

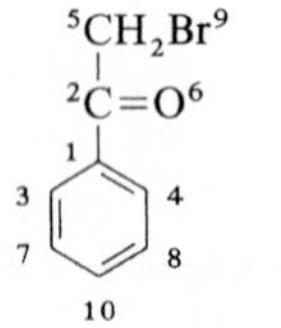

Node no.	Atom	Connected to	Bond type
1	C		
2	C	1	−1
3	C	1	*5
4	C	1	*5
5	C	2	−1
6	O	2	−2
7	C	3	*5
8	C	4	*5
9	Br	5	−1
10	C	7	*5
Ring closure		8−10	*5

1 = single bond   5 = alternating bond
2 = double bond   * = cyclic bond
− = non-cyclic bond

Figure 6. A nonredundant connection table (reproduced with permission from an internal publication of Chemical Abstracts Service)

mathematical term is "canonization". The Morgan algorithm used by Chemical Abstracts Service (CAS) was mentioned in Section 2.2.2.

In most modern chemical information systems structures are entered graphically and the connection table is derived by the computer. Fragmentation codes can also be derived algorithmically from a connection table.

The great diversity of connection table formats has led to moves to establish standard connection table formats for the exchange of chemical structure data. The Standard Molecular Data (SMD) file format is one of these [10.22], [10.23].

10.2 Compound Registration

An essential function of a chemical information system is the ability to search a chemical structures file for the presence of a given compound, perhaps represented in the file in a different but equivalent form. Compound registration is the procedure by which new compounds and accompanying data are added to a structure file and associated data files, respectively. A structure file in which a single and unique record of each compound is maintained is known as a registry file. A unique identifier is usually assigned to each structure record in a registry file.

The CA Registry File was described in detail in Section 2.2.2. The CA Registry Number is used not only in CAS publications and services but also in data banks, primary journals, handbooks, and compound lists.

Registration of a new compound involves matching a canonical representation [usually a connection table, although a unique WLN (Section 10.1.3) was used in older systems] against others already in the registry system. A modification of the Morgan algorithm is used in many systems [10.24], [10.25].

Registration may take place batchwise or online. In a *batch* registration system a batched, holding file of new records is periodically sorted and matched, usually against a serial file structure. In an *online* registration system, direct access to the structure file is required every time a record needs matching, but file update is then immediate, i.e., the new structure is available for searching within seconds of registration.

Direct matching of connection tables is inefficient for large files and various methods have been employed to reduce the match to a small part of the file. A simple method is isomer sorting (i.e., partitioning using molecular formula) [10.26], [10.27]. Another quick method involves generating so-called "hash codes" for every structure record [10.28]–[10.30].

10.3 Techniques of Substructure Searching

[10.2], [10.4], [10.31]

Chemical structure diagrams can be regarded as topological graphs in which the atoms are nodes and the bonds are edges. The problem of substructure searching is that of matching two graphs, one for the query structure and one for the file structure. This type of mapping is called *subgraph isomorphism*. A "brute force" approach involves trying all possible permutations. In the worst case situation the time taken to match a query structure against those on file increases exponentially with the number of nodes. The subgraph isomorphism problem belongs to a class of mathematical problems known as NP-complete.

Because matching is so time-consuming much research on substructure searching has focused on the development of efficient and effective screening methodologies, i.e., eliminating large numbers of structures that could not possibly be matched so that only a small part of the registry file needs to be submitted to the atom-by-atom matching procedure.

Screening Systems. Molecular formula offers an obvious screening possibility: for example, if palladium compounds are not required there is no point in doing atom-by-atom search on connection tables containing palladium atoms. Unfortunately most organic compounds contain carbon, hydrogen, and a few other elements, for which molecular formula screens are of no use. Thus screening systems based on algorithmically generated structural fragments have been devised. Atom-centered, bond-centered, and ring descriptor fragments have been investigated [10.32]–[10.36]. The BASIC (Basel Information Center) group of companies in Switzerland has compiled a dictionary of fragments [10.37] which formed the basis of the original CAS ONLINE substructure search system on STN International [10.38].

The DARC (Description, Acquisition, Retrieval and Correlation) system [10.39] offered commercially by Questel also uses "limited environment" screens but these are open-ended rather than controlled by a dictionary. They are discussed in more detail in Section 10.5.5.

Atom-by-Atom Search. Once screening has been completed (typically reducing the file to 1% of its original size), the second, atom-by-atom stage has to be carried out for those molecules which matched the query at the first stage. Most systems use "backtracking" [10.40] which involves matching one pair of atoms, then trying to establish matches for their neighbors, and so on. If at any stage a match cannot be established, the last unsuccessful match is "undone" and another one is tried. Partitioning, or set reduction techniques have also been tried [10.41]–[10.43]. These reduce the number of structure atoms which can be mapped to each of the atoms in the query and the number of combinations which need to be tested.

"Relaxation" is a technique which has been used more recently [10.44]. It is thought that parallel computer hardware (see Sections 7.2.1 and 10.4.4) is needed for this technique to be implemented efficiently [10.45].

Tree-Structured Inverted Files. The computationally intensive part of the substructure search can be reduced by preprocessing the data base to produce a tree-structured set of inverted files. Data-base building becomes slower but searching is more rapid. Two systems which use this technique are the Chemical Information System (CIS) originally developed by the National Institutes of Health (NIH) and the Environmental Protection Agency (EPA) [10.46]–[10.51] and the Hierarchical Tree Substructure Search (HTSS) developed in Hungary [10.52]. The CIS system offers three search methods: a Ring Probe (based on ring systems present in the structure), a Fragment Probe (based on augmented atom fragments), and an atom-by-atom search. HTSS extends the tree search file in the data-base preprocessing in such a way that substructure search does not need an atom-by-atom stage.

10.4   Current Research in Substructure Searching

10.4.1   Generic Chemical Structure Handling [10.53], [10.54]

Generic chemical structures, also known as Markush structures, are most frequently found in chemical patents as a means of expressing families of substances (see Section 19.3.4). They pose special problems. A single generic structure may cover millions of individual substances. The structure diagram may contain complexities such as variable bond orders and attachment positions, and optional rings. The generic description may be a mixture of text and graphics and the text may contain generic expressions such as "aryl", and non-structural concepts such as "electron-withdrawing group". All these factors have meant that chemical patents were best handled by fragmentation codes until recently. Examples are the Derwent chemical fragmentation code (see Section 10.1.2) and IDC's GREMAS code (see Sections 10.1.2 and 20.4).

The Sheffield University Project. Research on generic structure handling has been carried out at the University of Sheffield (UK) since 1979 [10.55]–[10.57]. An input representation for generic structures (GENSAL) has been developed [10.58]. It has a context-free grammar similar to that of modern programming languages and can describe a large proportion of the generic structures found in patents. The GENSAL Interpreter program [10.59] checks that the GENSAL statements are syntactically and semantically correct, and then generates an internal representation of the structure, the Extended Connection Table Representation (ECTR) [10.60]. Generic nomenclature terms are intrepreted by a chemical structure grammar, TOPOGRAM [10.61]. This can be used both to generate atom-centered and bond-centered screens for the generic radicals and to determine whether a substituent group in the query structure belongs to the class of radicals described by a generic radical term in the file structures. It can also determine common membership between generic radical terms.

A two-stage search process has been developed. The first stage uses algorithmically generated screens [10.62]. Atom-by-atom searching of generic chemical structures is not possible. A relaxation algorithm is used, which, due to the fact that it is inherently parallel in nature, offers the potential for implementation on a parallel processing machine [10.62].

A reduced chemical graph approach has also been developed to describe and match the gross structural features of substances. Some of the nodes are collapsed together, forming a smaller, simpler graph which is quicker to search [10.63]–[10.65].

Graphics-Based Systems [10.54]. As well as providing access to the CAS Registry File for its member companies, IDC runs a patent service and recodes patents from Derwent's Chemical Patents Index using the GREMAS code. The GREMAS system has a particularly good capacity for handling generic concepts in storage and retrieval and gives high precision in searching compared with other fragmentation codes [10.7]. However, it is difficult to learn and expensive to run. IDC is now developing a system allowing automatic generation of GREMAS codes from generic structures input using Sheffield University's GENSAL and Hampden Data Services (HDS) graphics structure input program PsiGen (see Section 10.6.3).

Hoechst, one of IDC's members, has developed a program GREDIA to generate a GREMAS search strategy from graphical input of a query [10.66].

Derwent Publications, Questel, and the French Patent Office, INPI (Institut National de la Propriété Industrielle) have collaborated in a project to enhance the DARC software (see Section 10.5.5) to permit generic data-base searching [10.67]. The new Markush DARC system will replace Derwent's fragmentation-code-based system. The backfile of fragments is already searchable using a microcomputer program called TOPFRAG, which allows the searcher to draw a generic structure using HDS's PsiGen techniques. TOPFRAG converts the structure into Derwent codes and formulates the online search strategy so that the user need not learn the rules of the coding system.

The efficiency of Derwent's Chemical Code, IDC's GREMAS system, and Markush DARC has recently been compared [10.9].

CAS introduced MARPAT in 1990 for searching a data base of generic chemical structures [10.64], [10.68].

10.4.2   Substructure Searching in Files of Three-Dimensional Structures

Three-dimensional (3-D) data are obtained mainly from X-ray and neutron diffraction methods. The 3-D coordinates, and cell and symmetry information, can be used to calculate a variety of intra- and intermolecular geometrical parameters. There are now well over 100 000 crystal structures in the public domain [10.69]. The major data bases are the Protein Data Bank, the Cambridge Structural Database (CSD), the Inorganic Crystal Structure Database, and the Metals Crystallographic Data File (see Section 9.5). Quantum mechanics and molecular mechanics programs may be used to calculate 3-D data, but the above data bases contain experimentally measured values.

The topography of a chemical structure (i.e., the arrangement of the atoms and bonds in 3-D space) can be represented in a connection table by replacing the interatomic connections with interatomic distances. A facility for 3-D substructure searching is of particular interest in computer-aided drug design. The task is often referred to as *pharmacophoric pattern matching*, where a pharmacophoric pattern is the geometrical arrangement of atoms in 3-D space responsible for some observed biological activity [10.70]. The first 3-D program was Molpat [10.71]. Lederle Laboratories have an in-house system, 3DSEARCH [10.72].

Research at Sheffield University. Research has been carried out at Sheffield University into techniques for searching a data base of 3-D chemical structures, analogous to those used for 2-D structures [10.73]. As with 2-D substructure searching, an initial and rapid search is used to eliminate large numbers of molecules followed by a more computationally demanding search (this time geometric) to determine whether the precise query pattern is present. The screen search is based on screens which correspond to pairs of atoms together with a distance range [10.74]. A range of algorithms has been studied for the geometric search [10.75] and multiprocessor systems are being studied to increase the efficiency of 3-D substructure searching (see Section 10.4.4).

A related problem to 3-D pharmacophoric pattern searching is the identification of a pattern responsible for some biological activity. Work previously done on the identification of maximal common substructures (MCS) for 2-D structures (see Section 10.8) has been extended to 3-D structures [10.76], [10.77]. Thus the largest common substructure in a set of molecules possessing the activity of interest is likely to contain the pharmacophoric pattern. Algorithms for the identification of an MCS are very computationally demanding and may perhaps be made more efficient by the use of parallel processing (see Section 10.4.4).

Cambridge Crystallographic Data Center. The Cambridge Crystallographic Data Center's GSTAT program locates a fragment, calculates user-specified geometry, and selects fragments on the basis of limiting values supplied for any of the derived parameters. Even then, statistical analyses may be required to answer a 3-D query completely. Integration of 3-D searching with the 1-D and 2-D capabilities of the CSD program QUEST is being effected via a 1:1 graph matching of chemical and crystallographic connection tables, to be followed by a careful generation of 3-D screening mechanisms. An improved statistics package is also being developed. Work on 2-D and 3-D similarity searching has also been carried out.

MACCS-3D. Molecular Design Limited (MDL), the vendors of MACCS (see Section 10.5.7) have recently released MACCS-3D for the organized storage, searching, and retrieval of static 3-D molecular models and model-related data, including atom and atom-pair data.

Models in MACCS-3D are MACCS-II structures for which additional 3-D information (e.g., Cartesian coordinates and partial atomic charges) is stored. A 2-D structure may have 3-D models associated with it, each corresponding to a different conformation or shape. Each model is associated with a single structure and has its own model-related data that may be general to the model or specific to individual atoms or atom pairs.

MACCS-3D allows exact-match, geometric, and submodel searching of 3-D models with geometric constraints specified to a certain degree of tolerance. Thus the user can retrieve molecules not only with a specific chemical structure, but with a specific conformation as well.

The user sets constraints (distances, angles, dihedral angles, and spheres of exclusion) on user-defined geometric objects (points, lines, planes, centroids, and normal vectors).

The multicolored display in MACCS-3D allows the user to compare queries with retrieved conformations by superimposing the query on each model retrieved.

A number of data bases are supplied by MDL for use with the software. Tripos' CONCORD program (developed by PEARLMAN and coworkers at the University of Texas) [10.78] has been used to create 3-D versions of MACCS 2-D data bases.

ChemDBS-3D. Chemical Design Ltd. have developed ChemDBS-3D, which operates in close integration with their other Chem-X molecular modeling modules.

ChemDBS-3D searches 3-D structure data bases for compounds matching a modeled pharmacophore in some low-energy conformation. It uses a 3-D distance screening method which enables all low-energy conformers to be searched but requires only one set of coordinates to be stored. The screening rate is 100 000 compounds per second. The search query (spatial distribution of centers) is perceived automatically from a modeled pharmacophore. Centers include user-definable classes of atoms with some specific property (e.g., hydrogen donor). Classical substructure search and property searches are also available. The results of screening and searching are written in nonconformationally dependent answer sets. Answer sets may be combined using arithmetic and logical operators, used as the basis for further searches, or exported to other systems. Conformations that match the original query are regenerated and stored in the results data base.

Aladdin. Daylight Chemical Information Systems offers the Aladdin system originally developed at Abbott Laboratories [10.79]. Aladdin searches data bases of 3-D structures to find compounds that meet substructural and geometric criteria. In addition, searches can be based on the ability of compounds to fit into predefined binding sites. Geometric objects (points, lines, and planes) are constructed from the coordinates of atoms in appropriate substructure environments using GCL (GENIE control language). Any object can be generated from previously defined objects, for example, a point as the mean of several previously defined points.

Search queries in Aladdin specify the ranges of distances, angles defined by three points, torsion (dihedral) angles, and plane angles that the geometric objects must match to constitute hits. Output files include details of each test, hard copies of structures with hit atoms identified, and data that can be used to generate molecular graphics of retrieved structures. Molecular models can be viewed as stereo images.

Within a framework called Daymenus, Daylight Chemical Information Systems supply not only Aladdin but also THOR (Thesaurus-Oriented Retrieval) for storage and retrieval of 2-D structures and data; and MERLIN, which performs substructure, superstructure, and similarity searching. Structures are stored as unique SMILES notations.

10.4.3 Similarity Searching

Registration involves exact-match retrieval; substructure search involves partial-match retrieval. Another possibility is a best-match or nearest-neighbor retrieval algorithm, where those structures in the file most similar to an input target structure are identified. Usually, the fragment screens that characterize the target structure are compared with the screens for each of the structures in the data base in order to determine the similarity; the structures are then sorted in order of decreasing similarity [10.80], [10.81]. This system allows chemists to browse in a data base without having to specify a substructure search precisely; it also gives them control over the volume of output [10.82].

Similarity measures can be used not just for matching a single target structure against a data base, but also for the pairwise comparison of all of the molecules in a data base to generate a clustering of them. Clustering methods can be used to group substructure search output and to select compounds for biological testing in drug design programs [10.83], [10.84].

10.4.4 Use of Parallel Computer Hardware

A well-known example of the use of parallel computer hardware to increase search speeds is provided by the CAS ONLINE system (see Section 10.5.2) where the early version of the Registry file was partitioned across thirteen pairs of minicomputers. The screens were stored as sequential lists on half of the pairs of computers and hits from a screen search were transferred to the other halves of the pairs for atom-by-atom searching. The hardware architecture made up for slowness in the search software, and screens and new compounds could be easily added to the system. The software, however, was not transportable.

In the newer "search engine" system implemented in 1989 [10.85], eleven advanced minicomputers are used, two of which serve as search managers, while nine search portions of the file in parallel. The main difference is that the screens are now stored as inverted lists. Search is now much faster (< 1 min as opposed to 5 min) and the system is transportable to single-processor environments.

The Beilstein Institute and Softron have developed a different multiprocessor architecture for substructure searching [10.86].

Four possible architectures are recognized for parallel computers:

1) Single instruction stream, single data stream (SISD)
2) Single instruction stream, multiple data stream (SIMD)
3) Multiple instruction stream, single data stream (MISD)
4) Multiple instruction stream, multiple data stream (MIMD).

MISD is unrealizable in practice. SISD represents the serial architecture of most computers to date, in which individual instructions are executed in sequence on one data type at a time.

An SIMD computer, the Distributed Array Processor (DAP), has been used at Sheffield University in research in nearest neighbor (similarity) searching in document data bases [10.87], [10.88] and in clustering of chemical structures [10.89], [10.90]. Cluster analysis is a multivariate, statistical technique that allows automatic production of classifications, usually by calculating all of the interobject similarities, dissimilarities, or distances. The use of clustering in drug-screening programs and in postprocessing of substructure search output is mentioned in Section 10.4.3.

Researchers in Sheffield have also used the DAP in searching techniques for 3-D macromolecules from the Protein Data Bank [10.90]. Their system, Protein Online Substructure Searching–Ullman Method (POSSUM), allows substructure searches to be carried out for user-defined motifs (i.e., patterns of secondary structure elements in 3-D space) [10.91].

Willet's group [10.90] reports: "POSSUM makes use of the fact that the common helix and strand secondary structure elements are approximately linear repeating structures and that such an element can be described by a vector drawn along its linear axis. The set of vectors corresponding to the secondary structure elements in a protein or a query motif can then be used to describe the structure of that protein in 3-D space, with the relative orientation of the helix and strand elements being defined by the interline angles and distances. Proteins and motifs may be regarded as labeled graphs, with the nodes of the graph corresponding to the linear representation of the helices and strands, and the edges to the inter-line angles or distances. It is accordingly possible to use subgraph isomorphism algorithms for the detection of motifs in protein structures."

The parallelism in an SIMD system, such as the DAP, arises from the fact that different processors can execute a program on their own local data. This is also true of a MIMD system but, in addition, the different processors can execute their own programs simultaneously and communicate between themselves as required. Early reports simulated the use of MIMD processing for substructure searching [10.45], [10.92] and in maximal common substructure algorithms [10.93].

More recently researchers at Sheffield University have used the INMOS transputer for the implementation of chemical substructure searching [10.87], [10.90], [10.94]. This transputer is a microprocessor which can be linked to others in a network, and programmed using a special language, occam. A multiprocessing system can only be used efficiently if the task to be carried out can be broken down into subtasks which can be allocated to various processors. In the Sheffield research, a "farm" or "pool" of processors is used. The subgraph isomorphism process has been implemented in two different ways: data base parallelism and algorithm parallelism. The former attempts to increase the speed of searching by distributing the data base across the nodes of the farm; the latter attempts to increase the speed of individual searches.

10.5   Operational Substructure Search Systems

10.5.1   History

ERNST MEYER was working on chemical structures in the 1950s even before CROSSBOW, see Section 20.4. The ICI CROSSBOW system [10.15]–[10.18] was the first substructure search system for in-house data bases to become established worldwide. It used not only WLN but also fragment codes and connection tables derived from WLN (see Sections 10.1.3 and 10.1.4). The CROSSBOW connection table was bond-implicit [10.21] as opposed to the bond-explicit connection table used by CAS.

The earliest system which allowed substructure searching involving chemical connectivity input and structure display, was the NIH/EPA CIS (see Section 10.5.3). Its Structure and Nomenclature Search System (SANSS) eventually allowed access to a large range of public data bases. The system allowed access from teletype terminals and was not graphics-based.

The CA Registry file became substructure-searchable online in the early 1980s, first by means of the DARC (Description, Acquisition, Retrieval and Correlation) system (see Section 10.5.5) and, soon afterwards, in the CAS ONLINE Service (see Section 10.5.2). These were the first online graphics substructure search systems.

Chemical reaction systems such as LHASA and SECS (Section 11.6.7) had long used graphics [10.95] but it was some time before the first in-house, proprietary system appeared. This was Upjohn's Compound Information System, COUSIN [10.96], [10.97].

Neither COUSIN nor the CAS ONLINE Messenger software was portable or commercially available. From the early 1980s there was a big demand for user-friendly interactive access to in-house chemical structure data bases. The market leaders became MACCS (Section 10.5.7) marketed by Molecular Design Ltd (MDL). DARC in-house and OSAC (Organic Structures Accessed by Computer) from the ex-Leeds University team ORAC Ltd (see Section 10.5.8) appeared later.

In graphics systems of this type the user draws a 2-D structural diagram representing his substructure query. The system automatically performs the search, usually by algorithmically generating fragment codes, doing a screen search, and then doing an atom-by-atom search.

Many organizations acquiring these new systems already had structures files on chemical typewriters or encoded in WLN and wanted to generate MACCS data bases or DARC connection tables automatically. Interconversion thus became a theme of the mid-1980s. ELDER's DARING program has been widely used for this purpose [10.19]. Further software is required to convert the DARING connection tables to MACCS or other versions, and to generate structure coordinates needed for the graphics display [10.19].

Once many in-house, chemical structure data bases had been built, users began to realize that it was more efficient to use commercially available structure handling software for structures alone (or structures and a minimal amount of related property data) and to take advantage of data-base management systems to handle property data [10.98], [10.99]. The vendors of chemical structure handling systems now allow for interfaces to data base management systems.

By the mid-1980s the advent of the microcomputer had started to make a big impact on the world of chemical information [10.100], [10.101]. The problem was no longer the lack of systems and data bases but rather the proliferation of systems which could not be linked in a seamless manner. Commercial and technical factors cause severe limitations on users [10.102] both as regards drawing chemical structures and accessing files with these structures [10.102].

10.5.2   CAS ONLINE [10.38], [10.103]

The CA Registry system, which contained about 10×10^6 compounds by 1990, has been described in Sections 2.2.2 and 2.2.5.

A number of files under the generic title CAS ONLINE are available online on STN International (see Section 2.2.6). The system software, Messenger, includes chemical substructure search, text retrieval and numeric data searching facilities.

Chemical structures and Registry Numbers are held in the CA Registry file. There are four ways of searching the structures:

1) EXA (or exact) search retrieves the input structure and its stereoisomers, homopolymers, ions, radicals and isotopically labeled compounds.
2) FAM (or family) search retrieves the same structures as EXA, plus multicomponent compounds, copolymers, addition compounds, mixtures, and salts.
3) For SSS a range of possible substituents and bonds can be used in the input structure and a substructure search is performed.
4) CSS (closed substructure search) is akin to a family search with variable groups allowed. (The concept of closing substitution is discussed later.)

Structures can be input textually using alphanumeric characters or graphically using a graphics terminal or personal computer. A number of graphics front-end packages can be used (see Sections 8.2.7 and 10.6.4). STN Express is recommended and is marketed by STN. Detailed examples of search strategies have been published [10.104]–[10.106]. STN also supply microcomputer tutorial programs under the name STN Mentor [10.107]. A sample search over 5% of the file can be run at no cost (other than connect charges). This projects whether the search will succeed or whether it is too wide.

The search can be limited by using a Registry Number range, adding screens, choosing search terms (e.g., to exclude polymers), and using hydrogen counts. Every position in the molecule is considered free for substitution unless otherwise specified. Thus the exact or minimum number of hydrogen atoms desired at each position must be specified with HCOUNT.

Certain generic groups can be specified (e.g., CY for any cyclic group). Variable point of attachment is also possible.

The CA Registry System automatically identifies substances with several possible chemically equivalent representations such as tautomers or aromatic rings [10.108]. Certain explicit single and double bonds in such circumstances are replaced by normalized bonds, and the associated migrating tautomeric hydrogen is associated with groups of atoms rather than single atoms. Since the single–double bond patterns and specific migrating group locations have all been replaced by normalized data, all forms of the tautomeric structure lead to the same Registry structure record. Unfortunately some users find it hard to understand the CAS conventions, since they are not quite the same as text-book definitions. Keto–enol tautomerism, for example, is not recognized. STN Express takes care of this problem, making single bonds single or normalized and double bonds double or normalized, where a tautomeric or aromatic system is possible.

At the moment stereochemical discrimination in search is not possible but the addition of true stereochemistry to the Registry file is planned in the early 1990s.

The output from a CA Registry file search includes Registry Number, CA index name, molecular formula, up to 50 synonyms, structure diagram, and full bibliographic information for the 10 most recent references about the substance. The total number of references for the substance in the CA file and an indication of further references in the CAOLD file are also given.

Once a Registry file search is complete, it is possible to switch to the CA file for a further search and for retrieval of the corresponding bibliographic information. The CA file is the only online file to contain CA abstracts and is not licensed to other hosts. The abstract text is searchable and displayable from 1970 onwards and older abstracts will be gradually added.

Information prior to 1967 is supplied as abstract numbers in CAOLD, where most Registry Numbers for compounds cited between 1957 and 1967, and retrieved in the Registry file, can be searched.

There are three other files in the CAS ONLINE family: MARPAT for generic structures [10.64], [10.68]; CASREACT, a reaction index (Section 10.8); and CA previews, where information is initially entered prior to abstracting and indexing.

The CAS ONLINE search system has screening and atom-by-atom search stages. The screen set was originally developed by the BASIC group of companies in Switzerland [10.37], [10.38] based on earlier work done at CAS and Sheffield University. CAS added further generic screens and CAS ONLINE uses 12 highly posted screen types. The original CAS ONLINE search software was not as sophisticated as some, but speed was achieved by parallel processing. The newer "search engine" system is faster still (see Section 10.4.4).

Comparisons of CAS ONLINE with other systems have appeared in the literature [10.109]–[10.114].

A number of companies, (e.g., Kodak [10.115]) maintain their corporate data bases as private registries under the CAS ONLINE system.

In 1989 CAS and Questel agreed to explore ways of collaborating and linking CAS ONLINE and DARC services. They also acquired minority shareholdings in Hampden Data Services who produce microcomputer software (including STN Express) for CAS, Questel, Derwent, and other vendors.

10.5.3   CIS/SANSS [10.46]–[10.51]

This system was originally designed for teletype terminals but graphics input is now possible using a PC-based front-end SuperStructure (see Section 10.6.4). SANSS is the substructure searching module for a large number of data banks on CIS (see Sections 9.4.5 and 9.6.3). CIS gives direct links to toxicological and spectral data rather than to bibliographies.

The structures are held as CAS connection tables. Tautomeric and aromatic systems have normalized bonds. The system contains about 350 000 structures, with molecular formulae, names, and synonyms. Since this file is much smaller than the CA Registry file, wider, more generalized substructure searches can be carried out without fear of an excessive output of hits. Moreover all the compounds are "useful" in the sense that they occur in data banks of commercial or environmental importance.

Searches can be limited to the contents of particular data banks. The searcher may employ name fragments (user-defined), molecular formula fragments (including ranges of acceptable element counts), and molecular mass ranges, as well as a controlled vocabulary of functional group and ring system codes to generate a small set of compounds on which to perform an exact substructure search.

If a ring skeleton is known but the number and/or positions of heteroatoms in the ring and/or points of substitution is undefined, a generic search at the desired specificity level can be run.

Atom-centered fragments of a particular type can also be searched. Sets from any of these operations may be combined with Boolean logical operators.

By using a variety of simple textual commands, the user can create the substructural fragments (user-defined type) for an atom-by-atom substructure search of a previously formed set of compounds.

It is not possible to block sites on the substructure with hydrogen to prevent substitution at that atom. Variable atoms but not groups can be specified. Stereochemistry is not handled.

The output record for each compound contains names, molecular formula, structure, and a list of all the collections in which the compound appears. Bibliographic information is not part of the system but can be located by downloading retrieved Registry Numbers and transferring them to another online retrieval system for use as search terms.

A nested-tree structure-searching system is used which operates on the data base of connection tables and uses an inverted file organization to provide rapid search response (see Section 10.3).

Comparisons of SANSS and other systems have been published [10.111], [10.114], [10.116], [10.117].

An adaptation of SANSS has been incorporated into the Drug Information System of the National Cancer Institute in the United States [10.118], [10.119].

10.5.4   Cambridge Structural Data Base (CSD) System [10.120]

The data base supplied by the Cambridge Crystallographic Data Center (Section 9.5) is one of only a handful of sources [10.69] of experimental 3-D structural data and a unique source of such data for organocarbon compounds. It is thus an invaluable tool in computer-aided drug design and in the analysis of geometrical parameters to obtain new insights into structure and bonding.

The data base is supplied complete with searching software. The core program QUEST in Version 3 allows users to search bibliographic data, chemical connectivity, and selected numerical data within a single query. Tests of individual fields are combined via logical operators to form a complete question. Version 4 (released in 1989) has a graphics input and output package which replaces the alphanumeric coding of the search question. Queries are formulated using a fully interactive, menu-driven interface. Substructure, textual, and numeric searches may all be specified.

Menus allow generation of a substructure which may be edited, saved for future use, or combined with further bibliographic or numerical questions. The complete query is then passed to the data-base search software.

Mapping of the crystal connectivity onto the chemical connectivity is promised in Version 5 of CSD. Version 4 is available in both machine independent and VAX/VMS specific hardware versions. It operates on all Tektronix terminals. It can also be used on certain workstations and under the UNIX operating system.

The Cambridge Crystallographic Data Center also supplies software for searching the 3-D data in CSD (see Section 9.5), and this is integrated with the 1-D and 2-D capabilities of the QUEST program.

Comparison of CSD with CAS ONLINE and MACCS. It is instructive to compare CSD with the leading online and in-house systems, CAS ONLINE (plus STN Express, Section 10.5.2) and MACCS (Section 10.5.7), respectively. Molecular Design Limited produce a microcomputer package, ChemBase, but it is not truly a front end to MACCS in the way that STN Express works with CAS ONLINE, so MACCS alone is used for comparison here. However, ChemBase has many features similar to STN Express. MACCS and CSD are both minicomputer/mainframe based products. STN Express has the advantage of being able to use advanced microcomputer features such as drop-down menus. For simplicity CAS ONLINE and its STN Express front end are called STN in the following.

The CSD provides over 60 system templates (pre-drawn structures which the user can select and modify). In CSD and STN the templates are displayed graphically and selected using the mouse, crosswires, or other pointing device. In MACCS (at least until 1991) the user has to know the file name of each template and select by name.

All three systems allow users to save their own templates. They all permit freehand drawing and a multiplicity of ways for constructing structures from all the atoms of the periodic table. STN and CSD offer "Feldmann ring notation" (e.g., 66U6D5 specifies the steroid skeleton). CSD offers more structure-drawing facilities than MACCS but is not as user-friendly as STN.

CSD draws structures on a grid system, producing "neat" structures with consistent bond lengths. STN allows "rubber-banding" of bonds so that atoms can be placed where desired. MACCS has neither a grid nor "rubber-banding".

STN and MACCS allow a roughly drawn structure to be "cleaned". STN allows the cleaning to be "undone". CSD and STN allow a variable point of attachment for a group, MACCS does not. STN has a wide range of "shortcuts" (e.g., SO_3H may be selected from a menu and the system will interpret this into the full topology of the group for structure storage and search). CSD has a similar "groups menu". MACCS has no equivalent. STN allows a rectangular window to be drawn around a part of any structure so that operations may be performed in that window alone. CSD's window can be irregular in shape. MACCS has no windowing facility.

Once a search is complete in MACCS the hits are made into a list. Creation of data-base subsets for offspring searching in CSD is somewhat clumsy and wastes disk space. However, Boolean logic is used in search statements in CSD whereas MACCS only allows Boolean logic for hit lists. Subsetting in STN is a very recent feature (mid-1990).

CSD incorporates similarity searching software. Two different coefficients can be used, the output is ranked, and the coefficient value is displayed. There is no similarity search in standard MACCS, but an additional module can be purchased. STN does not offer similarity searching.

The CSD data base contains over 90000 structural diagrams. In the search output the substructural search fragment is highlighted for rapid visual interpretation; this feature is not offered in MACCS or STN.

An additional CSD module, BUILDER, will soon (1991) be offered to allow users to build their own data bases. MACCS is already a data-base-building package and data bases are commercially available for use with it. At the moment STN is for use only with public online data bases.

10.5.5   DARC [10.39], [10.121]–[10.126]

The DARC system was conceived by Dubois of the University of Paris in 1963. It is now marketed by Questel, a subsidiary of Télésystèmes.

Generic DARC is used to access four online data bases included in the CA Registry: EURECAS, POLYCAS, MINICAS, and UPCAS. EURECAS is the main file, excluding polymers, POLYCAS contains polymers, MINICAS is a sample file (1%

of EURECAS), and UPCAS contains the most recent monthly update. Abstracts and pre-1965 data are not available; they are mounted only on STN. Chemical Abstracts connection tables are converted into DARC code for searching online on the Questel host.

Text structure input of a substructure query is particularly easy but graphics entry is normally used. This can be done with an online, interactive menu system, or offline using DARC CHEMLINK (a PC-based program with menus identical to those used online) or using STN Express (see Section 10.6.4).

A parent molecule is first built, then graph modifiers (up to 20 different groups or atoms) are specified. Boolean logic is allowed. Offspring searching can be carried out as follows. A structure is drawn and submitted to the first-stage search. Further searches can be carried out on the hits from this fragment search without incurring additional fees. Hits can be viewed after a first-stage search, whereas on STN the hits at the first fragment stage cannot be displayed.

The search can be limited by isolating rings (e.g., benzene is acceptable but not naphthalene) and by specifying the minimum and maximum number of rings, non-hydrogen atoms, and components.

If an atom is open to substitution, a "free site" must be specified. In CAS ON-LINE it is assumed that substitution is allowed unless hydrogen atoms are specified.

Once a substructure search has been carried out, structures can be displayed and the Registry Numbers can be transferred to search bibliographic files on Questel Plus (the CA bibliographic file, the Merck Index, the Janssen catalog, and the Eurosyn-thèses data base of dangerous chemicals, EECDNG). Cross-file searching with INPI patents files and Derwent's WPI/WPIL is also possible. This is particularly useful, compared with CAS ONLINE, because WPI/WPIL are not available on STN. A memory command can be used to sort and rank on any desired field in a set of answers. Unfortunately a hit file from a Questel Plus data base or a WPI/WPIL search cannot be used to generate a query for searching EURECAS. On STN it will soon be possible to take the Registry Numbers out of CA file hit output and use them to search the CA Registry file.

The DARC screening system is based on an open-ended fragment code—Fragments Reduced to an Environment which is Limited (FRELs). Two concentric layers of atoms around a focus are described and topological fragments of up to eight atoms in a row are then made. The focus can be an atom or a bond: the more highly connected the focus, the more discriminating the FREL. A tree structure with growing specificity is created from the fragments. The screens are file dependent and optimal for the file. An advantage is gained by adding a few extra screens for items such as ring information. The nature of the FRELs means that small molecules are handled better in DARC than in CAS ONLINE and system limits are less likely to be exceeded. Very large data bases (millions of structures) can be handled on a single processor machine. Search times for the CA Registry on CAS ONLINE and on Questel DARC are comparable.

Comparisons of DARC with other systems have been published [10.109]–[10.114], [10.116], [10.117].

Markush DARC. Whereas Generic DARC allows generic queries to be put to a data base of specific structures, Markush DARC allows generic queries of a generic structure data base [10.8]. Data bases searchable using Markush DARC are WPIM (Derwent Chemical Patents Index since 1987; see Section 20.3) and MPHARM (an INPI file of U.S. and European pharmaceutical patents since 1978 and French pharmaceutical patents since 1961). The bibliographies of WPIM are in WPIL and those for MPHARM in PHARM, also mounted by Questel. MPHARM and PHARM form the INPI data base PHARMSEARCH.

Since generic structural expressions cannot be simply represented by connection tables, Markush DARC has descriptors, known as superatoms, to represent types of structural elements (e.g., HEA stands for heteroaryl). Since specific compounds are not coded in the data base (which covers the scope of the invention, not all specific possibilities), there is no link between generic structures containing superatoms and the specific compounds to which they correspond. Thus a search for a substructure containing a superatom retrieves only structures containing the corresponding superatom.

A full structure, substructure, or generic substructure query can be phrased on- or offline; the search is performed; structures are displayed with highlighting of the relevant part of each structure; and bibliographic files can then be cross-searched.

Markush DARC is a collaborative development of Questel, INPI, and Derwent Publications. Questel are also collaborating with the European Patent Office (EPO) in the EPO Query system, EPOQUE.

DARC-In-house. In addition to online searching systems, Questel markets software for in-house use. DARC In-house is similar to MACCS in its functions (Section 10.5.7) and handles stereochemistry. Query input and structure searching are similar to those in Generic DARC; hash codes are used for full-structure search. Features such as report generation and structure registration are not applicable to online DARC. The DARC Communication Modules [10.127], [10.128] allow links to data-base management systems [10.129] and links are possible to microcomputer software.

The DARC "tool kit" is a set of stand-alone modules for editing structures and queries; searching on structure, substructure, and molecular formula; displaying and printing data; and handling queries and sets of answers. Communication modules allow links to data-base management systems, text retrieval software, and other applications.

Links between DARC In-house and the reaction indexing system ORAC (see Section 10.8) have been demonstrated.

10.5.6 HTSS

HTSS was developed in Hungary [10.52], [10.130]. It is now marketed by ORAC in Leeds (UK) and is the substructure search software used for accessing the Beilstein data base on Maxwell Online.

An HTSS data base has a hierarchic tree structure, which is created from the connection tables of the individual chemical structures by classification of all the atoms in the data base according to the number of nearest neighbors, atom type, and other classifiers. The subtree concept enables the search to be restricted to portions of the data base. If the required chemical structure is in the data base, then the tree of this structure is embedded in the tree of all the molecules which form the data base. The substructure search involves a "tree-walk" from root to leaves via the branches of the relevant tree which corresponds to the query substructure. A substructure is found if a leaf can be reached for each atom in the structure during the tree-walk; if not, the structures in the data base most similar to those requested by the query are found, enabling simultaneous retrieval of related structures. The HTSS approach integrates the screening and atom-by-atom searching steps used in conventional substructure searching systems, thus eliminating the time-consuming atom-by-atom search.

The time taken to search the tree does not increase linearly with the size of the data base, as is the case with most other systems. HTSS should therefore be suitable for use with very large data bases. Hicks and Jochum have shown that it is faster than DARC in many cases but they only vouch for its performance on files of up to 600 000 structures [10.113]. Walking across the tree sometimes needs a large number of disk accesses, especially when there are many free sites on the query structure.

Another possible disadvantage is the amount of time needed to generate the tree, although structures can be added to the data base without having to rebuild the whole tree.

An advantage of HTSS is that it can be implemented in almost identical forms on machines ranging from small microcomputers [10.131] to large mainframes.

10.5.7   MACCS

MACCS is the Molecular ACCess System marketed since 1979 by Molecular Design Limited (MDL) for integrated chemical information management on mainframe and minicomputers [10.106], [10.132]–[10.140]. The current version (MACCS-II) is the most popular of the systems for handling in-house chemical data bases and is available for a variety of makes of computer and graphic devices. It is an interactive, menu-driven graphics system, with facilities for graphical input, storage, retrieval, display, and printing of molecular information.

Structures input graphically are checked for valence errors and can be registered into corporate data bases. Templates are available (by file name, not by structure) to aid the graphics construction of diagrams. All the atoms of the periodic table, charges, and isotopes can be handled. Six-membered rings with alternating double and single bonds are recognized as aromatic. An additional "substance module" or "Mod S" handles more complex aromatic systems and has sophisticated features for mixtures and polymers. In basic MACCS tautomers and stereoisomers are checked before a compound can be registered.

Substructure search queries are entered using another menu in which a range of permissible atoms or bonds can be specified. A selection of atoms can be excluded as substituents at any position. All positions are assumed to be open to substitution unless hydrogen atoms are specified. Stereochemical discrimination in search is possible.

After a search, structures can be viewed one at a time on the screen. Data are displayed on different screens but software is available to print forms containing both structures and data for each record on a data base.

Generic queries (i.e., those where a group R is attached to the query skeleton) can be handled with the additional "Power Search Module" which also offers similarity searching.

The system is based on connection tables. Input structures are converted to a unique, nonvariant SEMA name (Stereochemically Extended Morgan Algorithm) [10.141]. Full-structure search uses hash coding of the SEMA name. The substructure search system uses about 1000 fragments (called keys) in a set similar to that of CAS, but smaller. The screens are held as inverted lists. Atom-by-atom search follows key search. According to [10.113] the system is limited to files of fewer than 350 000 compounds.

MDL sell a range of data bases for use with MACCS (e.g., Prous Science Publishers' Drug Data Report and The Fine Chemicals Directory). Data-base interface modules can be purchased to allow links to data-base management systems and the Customization Module allows the user to automate routine procedures, create customized menus with graphics buttons, and optimize the system interface to the local environment.

MACCS-3D offers 3-D substructure search (see Section 10.4.2). MDL also sell a reaction indexing package REACCS (see Section 10.8) with molecule handling features similar to those of MACCS. A set of microcomputer programs, CPSS (Chemist's Personal Software Series), is also available and includes ChemBase whose features are similar to those of MACCS (see Section 10.6.3). ChemBase is more user-friendly than MACCS and has some extra structure-handling features. However, this means that the two systems are not totally compatible, although a common file (the MOL or SD file) can be exchanged between the two. ChemBase is designed for use only with small personal data bases.

MDL also sell software for molecular modeling (CHEMLAB-II) and structure–activity relationships (ADAPT) [10.142], [10.143].

10.5.8   OSAC

Organic Structures Accessed by Computer (OSAC) is marketed by ORAC of Leeds (UK) and bears similarities to their reaction indexing system ORAC (see Section 10.8). MACCS, DARC, and OSAC are compared in [10.144].

10.5.9   Softron Substructure Search System (S4)

This system was developed by the Beilstein Institut and Softron of Gräfelfing (FRG). All molecules are encoded by using each atom in turn as the starting atom. For a molecule of *n* atoms there are *n* connection tables. The codes, which are very compact, are sorted, stored, and indexed. With this highly redundant representation, all hits can be retrieved with just a few sequential reads, carried out on a part of the file determined by the index. The need for many time-consuming random accesses is eliminated and atom-by-atom search is often unnecessary. Stereochemical and tautomeric search are not yet available.

S4 is claimed to perform faster than any other available system and can handle very large data bases (millions of compounds) [10.113]. It is also very suitable for searching CD-ROM data bases, such as Beilstein's Current Facts in Chemistry, where random access must be minimized because access is slow [10.145].

S4 is a full structure and substructure searching module. The complete system as implemented by Beilstein also requires the MOLKICK structure display and query editor interface (see Section 10.6.4). S4 is used by the Beilstein Institute in-house and is operated by DIALOG for their Beilstein online data base. It is also available commercially as an in-house system.

10.5.10   Proprietary Systems

Upjohn's COUSIN system (see Section 10.5.1) was developed in-house before user-friendly, interactive, graphics-based systems such as MACCS and DARC were available. Most companies choose to integrate a commercial structure handling package (and usually also a data-base management system) into their in-house integrated systems, rather than writing their own substructure search system [10.99]. Exceptions are, however, Bayer's ReSy (Research System), Pfizer's SOCRATES [10.146], and Philip Morris's MIMS [10.147]. The BASIC and IDC companies have had access to Chemical Abstracts and patents data in-house for many years. The development of graphics front ends for IDC's GREMAS system—GENSAL and GREDIA—is described in Section 10.4.1.

Sandoz is developing an end-user interface comprising a directory, a graphics module, aids for formulating property searches, and a communication module [10.148]. The in-house directory contains Beilstein, CA, and Fine Chemicals Directory structures in a hash-coded form. Sandoz corporate data are also available. Full structure and limited generic structure searching are possible but not substructure search. The graphics module is DARC. The communication module must give access to STN in a simplified, cost-effective form acceptable to Sandoz end users. "Hedges" such as "preparation" or "toxicology" have been set up for them. The provision of a uniform, consistent, STN/in-house substructure search facility is Sandoz's next priority. The problem of obtaining a universal interface to corporate and public data concerns most of the chemical and pharmaceutical industry [10.102].

10.6 Chemical Structure Software for Microcomputers [10.149]

10.6.1 Graphics Terminal Emulation

Graphics communication software, also known as graphics terminal emulation software, was the first area in which PCs made an impact on chemical information. Until such packages became available the only way of using services such as CAS ONLINE or DARC in graphics mode was to use very expensive graphics terminals. PCs supplied a cheaper alternative and allowed new possibilities such as offline query formulation and downloading (see Section 8.2.7).

10.6.2 Scientific Word Processing and Structure Drawing Software [10.149]–[10.151]

A scientific manuscript may contain not only text but also mathematical equations, diagrams, and chemical structures. Until recently only the text could be word-processed. Chemical structures and diagrams had to be manually pasted into the final document. Scientific word processors now allow the editing of equations, chemical structures, and text.

There are three methods of creating diagrams:

1) The character- or font-based method
2) The mark-up method
3) A graphics interface

The majority of packages with full-function word processing use the first method. Atoms and bond types are represented as single characters or fonts and the user builds a structure one character at a time. Some of these packages are described as WYSIWYG (What You See is What You Get) because the image on the screen corresponds very closely to the final print. The mark-up approach uses in-line codes that describe the drawing as a set of mnemonic instructions. The advantage of this is that machine-independent ASCII files can be written. However, the packages cannot be described as WYSIWYG. In the graphics WYSIWGG approach, a true graphics interface is used and free-hand drawing is possible together with selection of items from menus using a mouse or other pointing device.

With many scientific word processors a different application is used for each image type (equations, structures etc.). Thus, a true compound document that can be manipulated with one editor is not produced. Very few of the chemical structure drawing packages "understand chemistry", store connection tables, and offer substructure searching. Those that do are considered in Section 10.6.3.

ChemDraw from Cambridge Scientific Computing is probably the most popular structure-drawing package on the grounds of ease of use and print quality. It is available only for Macintosh computers. Various systems (e.g., CAS ONLINE, DARC, Daylight, and ORAC) offer ChemDraw links of one sort or another.

Graphics from external programs or scientific instruments can be incorporated into a document in the form of bit-mapped images (see Sections 6.2 and 11.5.2). This method has two disadvantages: the images require large amounts of computer storage space and they cannot be edited because they can only be changed bit by bit (see Fig. 2, p. 46). However, for hard-copy information a bit-mapped image is the only option. For electronically available information, a second option is a format known as a graphics metafile. Metafile images can be rotated and scaled-up or scaled-down, but they are not truly computable. The ideal method of importing graphics data is by means of a standard file format. Several instrument manufacturers are trying to agree upon such a standard.

10.6.3   Structure Management Software [10.149], [10.152]

Structure management software differs from structure drawing software in that exact structure and substructure searches are possible because connection tables are stored. Some of the programs also store and retrieve related text and numeric data. Structure management software allows users to build personal data bases of structures and related data. Compounds can be input graphically and registered into a data base. Records can be retrieved by structure or substructure search and hits may be displayed and printed.

ChemBase [10.149], [10.152]–[10.154]. ChemBase is one of the four modules of MDL's CPSS software for IBM-PCs. The others are ChemText, for structure drawing and scientific word processing; ChemTalk Plus, a PC-based communications package (see Section 10.6.4); and ChemHost, a mainframe or minicomputer program allowing corporate MACCS and REACCS data bases to be linked to ChemBase ones. ChemBase uses the same molecule editor as ChemText and presentation-quality graphics output are possible from both.

An initial data base of structures and reactions is provided as well as a large collection of graphically displayed templates. A mouse is required for graphics input. A variety of bond types are available including in-the-plane and out-of-plane bonds for designating stereochemistry. Valence checking takes place as the structure is drawn. Molecular formula and molecular mass are automatically generated. Substructure search is rather slow and has no screening stage, so ChemBase is suitable for data bases of only a few thousand structures. ChemBase is the only structure management system to handle reaction data bases and reaction-specific searches.

PsiBase [10.149], [10.152], [10.155], [10.156]. PsiBase (marketed by Hampden Data Services) incorporates a module, PsiGen, for creating, editing, and displaying chemical structure diagrams using drop-down menus and a mouse. PsiBase, including software for structure management and substructure searching, is available for both IBM and Macintosh microcomputers.

With PsiGen the user can draw structures freehand, build them from graphically displayable templates, or construct them using Feldmann notation. Commands can be entered by either menu buttons or keyboard commands. A variety of bond types are available including in-the-plane and out-of-plane bonds for designating stereochemistry. Parts of a structure can be enclosed in a rectangular "window" and operated upon independently from the rest of the structure. Valency checking takes place as structures are drawn. The molecular formula of a registered molecule is calculated.

Substructure searching is substantially faster than in ChemBase. Chemical reactions may be drawn, but not searched.

PsiGen is the structure drawing interface adopted by CAS for STN Express (see Sections 10.5.2 and 10.6.4), by Derwent Publications for TOPFRAG and TORC (see Section 10.6.4), and by Sadtler Laboratories for their spectral search system (see Section 9.4.5).

Other Structure Management Systems. HTSS (see Section 10.5.6) is available for use on microcomputers. PC-SABRE/PICASSO, from Fraser Williams (Scientific Systems) provides facilities for input, registration, storage, search, retrieval, and display of chemical structures and associated data. It also permits presentation-quality graphics output.

10.6.4 Special Application Software

Conversion of Connection Tables to Structures. ARGOS (Automatically Represents Graphics of Chemical Structures) is a program sold by Springer Verlag to convert ASCII connection table files into structures and to display them including stereochemistry, charges, and radicals. Fraser Williams (Scientific Systems) sell a program, PC-REWARD, which generates 2-D structure diagram coordinates from connection tables such as those produced by their program PC-DARING (see Section 10.1.3). The diagrams can be transferred to the PC-SABRE search package (see Section 10.6.3).

Property Prediction. TopKat from Health Designs Inc. is a menu-driven software package for the prediction of toxicity of structure input using the SMILES notation (see Section 10.1.3). It predicts rat oral LD_{50}, mutagenicity (Ames test), carcinogenicity, teratogenicity, and rabbit skin and eye irritation.

CompuDrug of Budapest and Austin, Texas sell Metabolexpert to predict metabolites of potential drugs. A related program, HazardExpert estimates health hazard effects in seven different biological systems, using the TSCA data base. Both programs allow the user to enter a chemical structure graphically.

Front Ends [10.157], [10.158]. Chemical structure searching presents a need for customized front ends, allowing the scientist to make use of the 2-D chemical structure diagram. When a user draws a query in the CAS ONLINE or DARC systems, information commands are sent through the telecommunications network to the central computer, which in turn interprets the commands and creates the query image for display at the user's terminal [10.101].

There are tremendous advantages to be gained if the diagram can be created at the local computer. This reduces the load on the central computer, reduces telecommunications traffic, and reduces the cost to the user. The stress caused by trying to remember all the commands that are required to create the diagram while the costs are ticking away (taxi-meter syndrome) is also reduced. Using an offline query negotiation, the user can think about the query and ensure that it is right before logging onto the online host.

In some of the packages which came on the market in the late 1980s, the writer of the front-end software (e.g., STN Express and DARC Chemlink) and the online host have collaborated closely, so that the best features of both systems can be used. The host must define standard formats for queries and also provide the appropriate software "hooks" in its retrieval system. However, products of this type tend to be exclusive: the front end is ideally used for only one host.

Other products (e.g., ChemTalk Plus) have been written without the collaboration of an online host, and are therefore prone to malfunction should the online system change minor features of its file structure or command language. These products depend on conversion of the input graphics structure to the text string used by the host. They cannot produce some of the "finer" features available in a collaborative venture.

ChemConnection is a Macintosh desk accessory written by SoftShell International Limited. Structures drawn with the ChemIntosh interface can be placed on the clipboard and converted to the text structure input string required for searching the STN Registry file. Uploading requires a communications program such as Versa Term Pro.

ChemTalk Plus is a module of MDL's CPSS software (see Section 10.6.3) for IBM-PCs. It provides terminal emulation facilities for online interactive access to MDL mini/mainframe software and (in conjunction with ChemHost) allows file transfer, structure and data searching, and downloading between MACCS and ChemBase. It also provides facilities for accessing public services such as CAS ONLINE and Beilstein Online. Offline query preparation, automatic logon, uploading, query execution, and display of results are permitted. For Beilstein Online (but not CAS ONLINE), the Chemists' Access System (a development of ChemTalk Plus) allows downloading of structures to make a substructure-searchable ChemBase data base.

DARC CHEMLINK is an IBM-PC-based package written to look as much as possible like DARC's standard chemical structure entry. It is exclusively for use with chemical structures on Télésystème's Questel.

MOLKICK [10.158], [10.159] is a package written by Softron for the Beilstein Institute. It uploads structures drawn with Beilstein–Softron graphics for searches of the CAS ONLINE Registry file on STN or Télésystèmes Questel, or for searching Beilstein Online on STN or DIALOG. Structures are converted into the CAS text structure input string or into a ROSDAL (Representation of Structure Diagram Arranged Linearly) string [10.158]. A function key allows rapid transfer in and out of MOLKICK and the online system. The program is quick, is memory-resident, and can be used with more than one terminal emulation package.

SANDRA (Structure and Reference Analyzer) is an IBM-PC program which takes the chemical structure of a compound and directs the user to where to find references to that compound in the Beilstein Handbook. The following information is output: the series, volume, and subvolume numbers; Hauptwerk page numbers (H-page); system number or range; degree of unsaturation; and carbon number (see Section 3.3.2).

STN Express is front-end software that provides access to STN International structure data bases (e.g., the CA Registry file) [10.106]. It is also a front end to Télésystèmes Questel. Features of the program include guided search, offline chemical structure query formulation, and offline search and strategy formulation. Help is supplied with tautomerism and other details. The program has special search-and-display features, data capture, and automatic logon. The guided-search feature allows the user to input a search query through a series of menus and then have it automatically processed in the STN data base. Offline chemical structure building is menu-driven and requires the use of a mouse. The structure input software is compatible with PsiGen (Section 10.6.3). The program provides predefined search strategies for general subjects (e.g., toxicology) that take advantage of data bases provided by STN. Graphics are fully integrated with text and can be captured in transcript files and printed. Versions are available for both Macintosh and IBM-PC.

SuperStructure is the Fein Marquart/Fraser Williams graphics front end for CIS (see Section 10.5.3). It evolved from the microcomputer-based structure input program written for the National Cancer Institute's Drug Information System [10.119], [10.120].

TOPFRAG converts chemical structures drawn on the PC to the correct Derwent chemical fragmentation codes, and compiles them into fully time-ranged search strategies for direct input to Derwent data bases on ORBIT, DIALOG, or Questel. Automatic conversions include tautomer perception and allocation of ring index numbers. The program allows the selection, through a menu-driven interface, of nonstructural concepts such as activities or uses. It also allows the user to edit input strategies. TOPFRAG handles generic structures and uses the PsiGen structure-drawing interface (see Section 10.6.3).

An equivalent program, TORC, (To RingCode) also uses PsiGen, but generates codes for searching Derwent Publications' data base of the pharmaceutical literature, Ringdoc.

Miscellaneous. Fraser Williams (Scientific Systems) sell a program, Markout, which accepts a series of structures, calculates a generic skeleton structure, and tabulates the various R groups on that structure in a matrix of R groups and data. Sorting on various fields is possible and the matrix can be used in structure–activity studies.

CASKit from DH Limited (Santa Cruz, California) can be used to capture graphics files from CAS ONLINE and process them into a form suitable for use with CPSS software.

10.6.5   Software for 3-D Molecular Graphics and Modeling

Three-dimensional drawing programs may serve as front ends to software on more powerful machines, but are basically limited to displaying and manipulating structures. Molecular modeling programs can calculate and predict a variety of physicochemical features of the drawn molecule. They may also create files compatible with modeling systems for use on more powerful machines. Modeling programs include an energy minimization function, through molecular orbital and/or molecular mechanics calculations.

The structure input features of both 3-D drawing and modeling programs are often less sophisticated than those offered by structure drawing and structure management programs (see Sections 10.6.2 and 10.6.3). However they offer more display options (e.g., stick, ball-and-stick, and space-filling displays). Some programs can also generate an ORTEP (Oak Ridge thermal ellipsoid plot) display. Rotation of the 3-D structure is also a feature. Hard-copy output is usually feasible.

Alchemy (Tripos Associates) is a well-known example for molecular modeling on an IBM-PC. It can be linked to the CAS Registry file via STN Express. CONCORD (see Section 10.4.2) has been used to generate 3-D coordinates for more than 4×10^6 organic substances in the CAS Registry. A 3-D molecular model can be uploaded from Alchemy through STN Express so that a search can be conducted in the Registry file. Literature references are thus found. The 3-D coordinates of interesting substances can then be downloaded through STN Express into Alchemy, ready for modeling and manipulation.

10.7   Structure–Activity Relationships and Drug Design [10.160]–[10.163]

Quantitative structure–activity relationship (QSAR) techniques involve the correlation of physical, biological, and chemical data with variables characterizing the molecular structures in a data set. Most of the applications have been in pharmaceutical and agrochemical research because of the costs involved in synthesis and testing of thousands of compounds. The methods are, however, applicable in other areas, such as environmental chemistry [10.164]. In an agrochemical or pharmaceutical synthetic program, SAR methods may be used for lead optimization or lead generation. Lead optimization attempts to optimize activity within a given series of compounds by systematic modification of compounds previously identified as active. Lead generation attempts to identify classes of interesting compounds that might be active.

Various property descriptors have been used: hydrophobicity, octanol–water partition coefficients, pK_a, electronic descriptors (Hammett sigma constants, dipole moments, molar refractivities, ionization potentials). These are correlated with 2-D substructural fragments (e.g., augmented atoms, topological torsions) or 3-D substructural fragments (e.g., pharmacophores).

Techniques may be classified as statistical approaches (Section 10.7.1); substructural analysis and data-base techniques (Section 10.7.2); and molecular modeling approaches (Section 10.7.3). Molecular modeling and quantum mechanics can only be used for small numbers of structures; statistical methods can handle hundreds of compounds; substructural analysis can handle thousands.

10.7.1 Statistical Approaches

Free–Wilson Additivity Model [10.165]. This approach does not use physicochemical parameters but assumes that each constituent in a given location makes an additive and constant contribution to the overall activity of the molecule. A series of equations can then be written expressing the biological activity of the molecules in a set in terms of a constant activity plus contributions from each substituent. The equations are solved by regression analysis. The modification used by FUJITA and BAN overcomes the problem of the need for multiple substitution sites [10.166].

Hansch Analysis [10.167], [10.168]. This involves the correlation of physicochemical properties with the observed biological activities within congeneric series of compounds, mainly using multiple regression analysis. HANSCH combined several properties into a multiparameter model in which the biological response is modeled in terms of electronic, steric, dispersion, and hydrophobic components. HAMMETT and TAFT's earlier work [10.169] formed the basis of the electronic and steric contributions; the dispersion component is represented by the molar refractivity and the most popular hydrophobic parameter is the octanol–water partition coefficient. The success of the model has led to the inclusion of many other parameters [10.169]–[10.172]. However, the method is usually limited to structural variations within a given congeneric series where there is a high probability that the same mode of action pertains for all of the compounds.

Pattern Recognition. Pattern recognition methods were first used in chemistry for structure elucidation in the field of spectral information [10.173] but this led to studies involving predictions of biological activity [10.143], [10.174].

An example of a pattern recognition technique is a linear learning machine, where compounds are considered as points in multidimensional space, each dimension of which corresponds to one of the structural features used in the characterization of the compounds in the data set. The machine tries to identify a hyperplane dividing the space so that all of the active compounds lie on one side of the plane. Other pattern recognition techniques have been used [10.143], [10.160]. An interactive program, ADAPT, based on pattern recognition techniques [10.142], [10.143] can be used in conjunction with MDL's MACCS software (see Section 10.5.7).

Cluster Analysis. Unlike pattern recognition, which categorizes compounds into predefined classes, cluster analysis attempts to detect groups present in multivariate data sets in order to identify classes. Clustering in chemical information is described in [10.83].

10.7.2   Substructural Analysis and Data-Base Techniques

The Free–Wilson approach (see Section 10.7.1) forms the basis of the lead generation technique known as substructural analysis in which a given substructure is assumed to make a constant and additive contribution to the overall activity of a molecule, irrespective of the type of molecule in which it occurs. Structural features which can be derived automatically from computer-readable structure representations, and used in QSAR and property prediction studies are described in [10.175].

The first approach to substructural analysis used features derived from a fragmentation code [10.176]. The approach has been extended and refined and is referred to as the statistical–heuristic method [10.177], [10.178]. It has been used in the selection of compounds for antitumor screening at the National Cancer Institute. Compounds are ranked in order of decreasing probability of activity. MEYER's work with fragment structures is also notable [10.179].

A multiple regression analysis technique uses both WLN and connection tables as representations from which structural features can be automatically derived [10.180]–[10.186].

The use of data-base techniques such as 3-D searching was described in Section 10.4.2.

10.7.3   Molecular Modeling [10.187]–[10.192]

Physical models of molecules are of limited value, especially for large molecules. They may take a long time to build and can be easily damaged. They are also difficult to manipulate and represent a static molecule without indication of its energy. Molecular graphics is a much more useful aid in visualizing 3-D structure. It combines interactive computer graphics with computational procedures such as molecular mechanics or molecular orbital calculations [10.193], [10.194].

A number of molecular modeling systems are commercially available (e.g., Chem-X from Chemical Design, QUANTA from Polygen, and SYBYL from Tripos Associates). They enable the user to manipulate objects in real-time on the screen, rotate them, translate them, and scale them to any size.

The 3-D shape and structure of a molecule are important to its biological activity and molecular graphics techniques are increasingly being used in the field of drug design. They allow 3-D interactions to be viewed between a potential drug molecule and its biological receptor site.

If crystal structure data for a potential new drug are not available, the 3-D structures can be produced from calculations based on standard bond lengths, bond angles, and dihedral angles. The 2-D structure is drawn and input to a molecular mechanics energy minimization program. A distance geometry approach and quantum mechanical calculations are other methodologies for obtaining preferred conformations.

Molecular graphics systems can be used to study interactions as a substrate molecule approaches the site of biological action. Using a technique known as plane clipping, an image can be moved in a direction perpendicular to the plane of the screen so that the molecular surface can be viewed from points both inside and outside the molecule. Techniques for enhancing 3-D perspective include the shading and color-highlighting of atoms and bonds, variation in the intensity of illumination (known as depth cueing), and the removal of "hidden" lines.

Molecular graphics can be used to identify pharmacophores (see Section 10.4.2).

10.8 Reaction Indexing [10.195]

Two major problems need to be overcome before retrieval systems for reactions become as well-established as systems for storage and retrieval of individual molecules. The first is the very wide range of query types that has to be handled. These include reactions where both reactant and product are fully specified; reactions giving rise to a specified structure or substructure; reactions of a specified structure or substructure; and a substructural transformation in which reactant and product substructures are specified.

The second problem is the organization of the large volumes of data that need to be stored to answer other types of query. Apart from topological descriptions of any chemical structures, a system must hold experimental conditions (temperature, pressure, yield, etc), citation details, and a reaction analysis (i.e., some description of the change that has taken place). This last is the most important, and most difficult element.

VLADUTZ suggested that reactions could be indexed by substructural changes, and introduced the concept of a reaction site [10.196]. This consists of all the bonds altered during the reaction, including any heteroatoms directly connected to an atom in the reaction site and any atom connected by a multiple bond to the site.

The automatic identification of such reaction sites involves comparing connection tables to identify common features on two sides of a reaction equation [10.195], [10.197], [10.198].

Operational reaction indexing systems such as GREMAS (Sections 10.1.2 and 20.4) and CRDS (Section 3.4) have been considered elsewhere. The present section relates to graphics-based systems for storage and retrieval of chemical reaction information.

CASREACT [10.199], [10.200]. CASREACT is one of the CAS ONLINE data bases available on STN International (Section 10.5.2). Indexing for the service began in January 1985. At the moment only reactions in ca. 100 journals covered in CA's Organic Section are included in the data base, patents are not included. By 1988 the data base size had grown to 625 000 single-step reactions; ca. 170 000 are added annually. The system is document-based rather than reaction-based. The total number of matching reactions in an answer set of two or more documents cannot be identified.

Substructure searching is done in the CA Registry file. The whole file is not searched because the subset of compounds relevant to CASREACT can be retrieved by use of a screening fragment. Once the Registry file search is complete the hits are crossed over into CASREACT and associated with a "rôle" (e.g., "reactant").

The display of hit reactions consists of the reaction "map" ($A + B \rightarrow C + D$ for example), followed by a reaction scheme of chemical structures and Registry Numbers for the reactants, products, reagents, catalysts, and solvents. Abbreviated names are given for common catalysts, reagents, and solvents; a full CAS Index Name is given for unusual molecules. Only the Registry Numbers are searchable, not the names or abbreviated names. A Note (NTE) statement comments on hazards, unusual conditions (e.g., photolysis), and failed reactions. The final part of the display is the bibliography, which is not searchable in CASREACT but can be searched by crossing into CA File.

A comparison of CASREACT and REACCS has been published [10.201].

Organic Reactions Accessed by Computer (ORAC) [10.202]–[10.205]. ORAC is a graphics-based reaction indexing system written at Leeds University and sold by ORAC Ltd of Leeds. The related system for handling chemical structure data bases, OSAC, was described in Section 10.5.8. Searchable features in ORAC include reactants, products, intermediates, reagents, yield, reaction conditions, author, substructure, and reactions. The reaction search allows retrieval on the basis of reaction site or functional group changes. Keywords are also used for searching and retrieval because they can convey concepts which are difficult to express in terms of structural features (e.g., mechanism, selectivity, or reagent types).

Up to 32 "boxes" (data bases) can be handled. These include in-house data bases and the ORAC data base of over 116 000 reactions (1990 figure) from the current literature and key tertiary publications. They can be searched separately or in any combination.

There is a similarity searching feature. A Host Language Interface allows links to other systems both for synthesis planning and for QSAR and modeling [10.206].

REACCS [10.106], [10.204], [10.207]–[10.209]. REACCS is MDL's Reaction Access system and is related to MACCS (see Section 10.5.7). It is an interactive graphical data-base management system for storing and retrieving chemical reaction information. Reactants, products, catalysts, reagents, and solvents are substructure-searchable, with graphics input and output of the diagrams. Reaction centers are perceived. Variations on a given reaction are grouped together. Thus reactions are not duplicated if they differ only in yield or experimental conditions. Stereochemistry can be stored and searched. Multistep reactions are indexed by making manual entries for each individual step and the overall step. Similarity searching is offered.

One of the reasons for REACCS' prominence is the availability of several large and useful data bases with the system. These include a Current Literature File (CLF), the Theilheimer data base, and the Journal of Synthetic Methods data (see Section 3.4), a REACCS-readable version of Organic Synthesis, METALYSIS (a data base of metal-mediated chemistry), and CHIRAS (a data base of asymmetric synthesis).

Other Reaction Indexing Systems. SYNLIB [10.204], [10.205], [10.210] differs from REACCS and ORAC in that it could be classed as a knowledge-based system. It was designed to facilitate browsing in an evaluated collection of representative reactions. Users can also build their own in-house data bases and modify the supplied knowledge base. Comparisons of SYNLIB, ORAC, and REACCS have been published [10.204], [10.205].

Pfizer have written their own system, CONTRAST [10.211], as part of their in-house system SOCRATES [10.146].

10.9   Computer-Aided Synthesis Design

Programs for computer-aided synthesis design may be knowledge-based or depend on physical parameters. In the latter structural transformations are represented in a generalized manner by manipulating an abstract model of atoms and electrons.

LHASA, SECS, and CASP (see Section 11.6.7) are well-known knowledge-based systems. Other examples are CHIRON (Chiral Synthon) [10.212], and SYNCHEM [10.213], [10.214].

Systems based on physical parameters are EROS, IGOR, SYNGEN, and CAMEO. EROS (Elaboration of Reactions for Organic Synthesis) [10.215] uses DUGUNDJI and UGI's mathematical theory of constitutional chemistry [10.216].

IGOR (Interactive Generation of Organic Reactions) is based on the same model [10.217]. It has been used to identify novel reaction types [10.218].

SYNGEN aims to find the shortest and most economical synthetic routes to a given target structure from available starting materials [10.219], [10.220]. A rigorous mathematical form is used for describing reactions. Structures are represented numerically, enabling all possible reactions to be considered.

CAMEO (Computer Assisted Mechanistic Evaluation of Organic Reactions) is based on mechanistic analyses and operates in a synthetic, rather than retrosynthetic direction. It is an interactive program designed to predict the products of organic reactions from specified starting materials and conditions.

The AIPHOS system combines both approaches to synthesis planning [10.221].

11 Artificial Intelligence

11.1 Introduction

The goal of artificial intelligence (AI) is to construct computer programs which exhibit behavior that would be called "intelligent" when observed in humans. These programs seek to emulate human responses, for example, to solve problems, to learn from experience, to understand language and to interpret visual signs [11.1]–[11.9].

The roots of AI go back to 1936 when ALAN TURING first theorized about teaching computers to perform tasks based on logic. In 1950 he proposed what has come to be called the Turing test [11.10]. If a machine could fool its human partner into believing it to be human by participating in a wide range of conversations, then it could be concluded that the machine had passed the Turing test.

AI will be of even greater significance in the future because of its links with fifth-generation computers (see Chap. 6) [11.11], [11.12].

DAVIES defines the six classical branches of AI as machine architectures, expert systems (Section 11.6), natural language processing (Section 11.5.2), robotics, speech simulation and recognition (Section 11.5.1), and vision [11.13]. Robotics and computer vision (attempting to make machines identify and track objects they "see", and make decisions about them) are of little interest to the information profession at present. Optical character recognition (see Sections 6.2 and 11.5.2) is not usually regarded as part of vision technology.

CERCONE and McCALLA include as AI subareas: search, problem solving, and planning (Section 11.3); theorem proving and logic programming (Section 11.4); knowledge representation (Section 11.6.2); learning; computer-aided instruction; game playing; automatic programming; and AI tools [11.7].

Machine learning strives to develop methods for automating the acquisition of new information, new skills, and new ways of organizing existing information [11.14]. Automatic programming (getting the computer to program itself) is now considered to be a formal area of computer science separate from AI itself. Computer-aided instruction [11.15], game playing, and miscellaneous AI tools are beyond the scope of this book.

11.2 Machine Architectures and Neural Networks

Machine architecture work deals with fundamentally different designs for computers, e.g., parallel processing (Sections 6.1 and 10.4.4). Most new types of hardware require sophisticated new software.

Neural networks represent a new way of organizing the computation process and are often implemented by simulation rather than by building specialist hardware [11.16]. They are data processing systems that learn by example. Design is inspired by the way in which the brain works with a large number of single processing elements operating in parallel. Neural networks are most powerful when applied to problems whose solution requires knowledge that is difficult to specify. Possible applications are in image recognition, computer vision, speech processing, robotics, and knowledge processing.

11.3 Search, Problem Solving, and Planning

There are two basic ingredients in intelligent behavior: possession of knowledge and search. The organization of knowledge is considered in Section 11.6.2. Search refers to the ability to create a space of possibilities large enough to contain the solution to a problem, and then searching for that solution.

Attacking a problem involves moving from a starting state to a goal. In chess the starting state is the initial board layout and the goal is checkmate. The number of possible states from the starting state to the goal defines the state space, which for chess is about 1×10^{120}. Each move produces a branch path that grows towards a potential goal. As the number of possible pathways increases, a combinatorial explosion takes place. Combinatorial explosion can limit the capability of an intelligent program so other problem-solving techniques, such as problem reduction and heuristic search [11.7], have been developed.

11.4 Theorem Proving and Logic Programming

Theorem proving is the process of making logical deductions from a noncontradictory set of axioms specified in first-order logic or predicate calculus. The process can be automated using a method called resolution [11.7].

Many simple problem-solving tasks can be formulated and solved using a theorem-proving approach, but this has not been widely applicable because of the problems of combinatorial explosion. However, theorem proving is at the heart of the more recent development of logic programming. The AI language PROLOG (PROgramming in LOGic), developed in 1972 [11.17] is based on a resolution theorem prover (see Section 11.6.5).

11.5   Human–Computer Interaction [11.15], [11.18].

The user interface (the way information is entered or output) was discussed in Chapter 7. The use of user-hostile command languages is gradually giving way to user-friendly methods such as the selection of items from menus using pointing devices and touch screens. None of these are AI techniques. Much AI research has, however, been done on human–computer interaction (HCI), sometimes also called man–machine interaction (MMI), particularly in computational linguistics.

11.5.1   Speech Simulation and Recognition

Many of the issues of natural language understanding overlap those of speech understanding (see Section 11.5.2).

Speech as an output medium is well established. Early speech synthesizers digitized a speech signal for subsequent reproduction but more advanced systems generate the signal as required. Such synthesis methods are more flexible in operation and less demanding of computer storage, while producing more acceptable, natural speech output. Speech synthesis chips have been employed in toys, car and aircraft alarm systems, and automatic telephone answering.

The use of speech as an input medium is a more challenging problem [11.19]. Minimal acoustic differences between many words, background noise, and pronunciations cause problems. A Kurzweil machine can currently recognize a vocabulary of about 5000 words, when spoken as discrete units and not run together as in rapid human speech. A small vocabulary is sufficient for certain applications such as bibliographic retrieval [11.20], password or telephone number identification, air traffic control, and voice activation of domestic appliances or robots.

Connected speech recognition remains an unsolved problem. An example of the state-of-the-art is the HEARSAY-II system developed at Carnegie Mellon University [11.21].

11.5.2 Natural Language Processing (NLP)

Natural Language Interfaces. Rigid interfaces, protocols, command languages, and syntax requirements are serious barriers to the widespread use of computers. The user would prefer to express his requirements in his own language in a form such as "Give me a good 1988 review article about the treatment of asthma".

Unfortunately natural language is complex and ambiguous. A computer can easily compare character strings but it cannot detect the equivalence of "I hold a book" and "The volume is in my hands" [11.22].

Functional, relatively reliable, NLP systems are now available for several question-and-answer environments with a small possible vocabulary and a small number of queries and responses. An example is MYCIN, an early system for medical diagnosis [11.9], [11.22], [11.23]. The expert data base searching system TOME searcher [11.24]–[11.26] has a natural language interface for a domain-specific knowledge base in the field of information technology.

Understanding natural language involves three levels of interpretation: syntactic, semantic, and pragmatic [11.7], [11.27]. *Syntactic processes* parse sentences to clarify the grammatical relationships between words in a sentence. *Semantics* is concerned with assigning meaning to the various syntactic constituents. Semantic analysis converts the output of syntactic analysis into an extended predicate calculus, or a semantic network with quantification, and resolves ambiguities arising from the multiple meanings of words. *Pragmatics* attempts to relate individual sentences to one another and to their context.

Machine Translation [11.28]–[11.31] Machine translation has had a checkered history over the last 40 years and many translators still regard translation programs as "toys" of the academic researcher. The translation of technical manuals demands few cultural and stylistic skills and only a limited vocabulary may be required. Computer systems can be customized to handle such material reasonably well. Up to 80% of a human translator's time is spent in terminological research and document production. There is, therefore, considerable scope for "tools" such as electronic dictionaries linked to word processing software. The vendors of machine translation systems hope that the computer can relieve humans of the boring and repetitive elements of their tasks.

The Commission of the European Communities' machine translation system SYS-TRAN has been operational since 1976 and is available for several European language pairs. Translations are processed at a rate of up to 400 000 words per hour. The EUROTRA project has been running since 1982 and is designed for the nine official languages of the EEC (72 language pairs).

Most Japanese electronics companies are developing or selling machine translation systems, mainly to translate English into Japanese. NEC has a prototype system that translates between Chinese, English, Japanese, Korean, Spanish and Thai using a "Pivot Method" where an intermediate language is used as a go-between. Fujitsu has Japanese-to-English and Japanese-to-German systems for business contracts, technical documents, and manuals. Researchers at ATR Telephony Research Laboratories, in cooperation with the Center for Machine Translation at Carnegie Mellon

University, are applying neural network principles to speech recognition and automatic translation systems. They hope to mount a real-time, telephone translation system in Japanese–English/English–Japanese.

Optical Character Recognition (OCR). Progress is being made in computer recognition of written input (OCR). Sophisticated OCR machines (e.g., the Kurzweil) can be trained to recognize almost any print face. "Intelligent" optical character recognition systems are referred to as ICR, rather than OCR.

Scanning is done through charge coupled devices (CCDs) that transform light reflected from the image into signals that are a function of the light intensity. The image is then scanned and broken down into minute picture elements (pixels) that are sent to a computer where the digital files are created. The files are then processed by software for a particular application. The image and application files can be compressed and decompressed to more efficient sizes.

Several approaches can be used to translate the pixel bits to the multibit bytes used to define word characters. Feature extraction is based on the principle that each character has distinct features, regardless of font or spacing considerations. The software analyzes the scanned character, builds a features list, and then determines which character has most or all of these features. Some more advanced systems use AI techniques and can learn the characteristics of a new font. Some systems even attempt to read handwriting.

Neural network technology promises further breakthroughs in the reading area.

An example of the practical use of OCR in chemical information is the way CAS converted their older indexes into machine-readable form (see Section 2.2.5).

11.6   Expert Systems [11.8], [11.32]–[11.40]

11.6.1   Definition and Features

An expert system handles real-world, complex problems requiring an expert's interpretation. It solves these problems using a computer model of human reasoning [11.36]. Expert systems are the most important component of research in intelligent knowledge-based systems (IKBS), which are in turn a component of fifth-generation systems (see Chap. 6) [11.11], [11.12].

Artificial intelligence and expert systems differ from traditional computing in three ways:

1) They work with symbols rather than numbers.
2) They reason with heuristics ("rules of thumb") rather than algorithms (precisely defined instructions and decisions). The heuristic method of problem solving involves trying potential solutions, evaluating the result, and then modifying the procedure.

3) AI and expert systems work with interpretative rather than compiled languages, allowing the expression of concepts difficult to encode in traditional languages. Problems expressed in AI languages are transformed directly into machine actions during run time. However, compiled programs execute much more rapidly and efficiently than interpreted ones.

Expertise is acquired and codified during interviews with experts in the relevant problem area (domain). It is codified as a collection of facts and rules that constitute the *knowledge base*. Some systems are able to handle uncertain information using numerical values to denote the degree of confidence, credibility, or plausibility. The knowledge base is controlled and operated by the *inference engine*. The inference engine is separate from the knowledge base, thus allowing the knowledge base to be updated and extended without interfering with the overall control structure. In many expert systems interaction with the system occurs through a *user interface*, not directly with the knowledge base. The system should be interactive in that it can ask the user questions when the information it has is incomplete or conflicting.

11.6.2 Knowledge Representation [11.4], [11.32]

For a program to exhibit intelligence it must have access to large amounts of knowledge and it must know how to manipulate and use that knowledge [11.37].

Knowledge must be represented inside an expert system so that it can be used effectively, modified, and augmented. Representation must be accessible and flexible. The expert system must also be able to explain its actions to humans.

A number of different knowledge representation "paradigms" have emerged. One of the earliest represented knowledge in *semantic networks* where facts are stored at nodes and relationships between the facts are represented by arcs. One of the most common relationships in a semantic network is the so-called "ISA" link which allows facts (e.g., dogs have tails) to be attached to classes of objects (e.g., dogs) and then inherited by specific objects in the class (e.g., Fido, Rover).

Each node in a semantic net has to be meticulously specified. Generic objects are not handled efficiently; difficulties are encountered in interconnecting object attributes and in handling unspecified attributes or default conditions. Expert systems developers therefore started to study *frames* for the storage of declarative knowledge. Here data are presented in a hierarchy of facts where lower orders automatically inherit all the characteristics of higher orders. Thus a specific breed of dog must exhibit all the features of dogs in general. Frames are composed of slots. For example, a frame for a chemical could contain slots for formula, structure, and physical properties. Slots may be further subdivided into facets. Slots may also contain hypotheses that relate to the expert system's functioning, rules about program situations, or pointers to other frames. Frames allow an expert system to retain more information about a specific item than it must explicitly express each time. Quantities of diverse information can be conveniently handled and extra slots can be added for new data.

First-order predicate calculus represents information by means of formulae. Function, variable, and constant symbols are set off with parentheses, brackets, and commas, respectively, to form statements. True or false propositions express relationships between specific and generic objects. More complex expressions can be written using connectives such as logical AND, logical OR, "implies", and "there exists".

The knowledge base not only stores declarative knowledge. It also tells how the data can be manipulated to solve the problem (procedural knowledge). The most common method of data manipulation involves production rules of the type: IF (condition) THEN (action).

The condition and the action often have several components, e.g., IF the unknown is a bird AND the unknown cannot fly AND the unknown has black and white feathers THEN it is an ostrich with probability 0.8. Predicate logic and other techniques can also be used to represent procedural knowledge.

11.6.3   Knowledge Engineering

Domain knowledge has to be elicited from the human expert by a "knowledge engineer" using interviewing and observation techniques. The knowledge engineer has to structure the knowledge so that it can be properly represented by the computer system. As knowledge engineering is slow, methods are being been sought to automate knowledge acquisition or allow systems to learn from examples presented in a natural way.

Computerized tools have been designed to help the knowledge engineer (see Section 11.6.5). Many personal-computer-based expert system shells and some larger, more complex expert system building environments incorporate more reasoning strategies, and have choices of representation language, graphics, and links to other languages.

11.6.4 Inference Engine [11.34], [11.41], [11.42]

The inference engine (control structure) is the central program which manipulates the data in the knowledge base to reach conclusions. The engine may approach a problem by beginning with either hypotheses or facts. A *goal-directed expert system* uses a reverse-chaining or backward-chaining algorithm. The program starts with a limited set of possible hypotheses and attempts to prove the validity of each one by examining all the factors. The knowledge base is searched to find a rule which concludes the initial hypothesis. The IF clauses from this rule then become the hypotheses for the next search level. The process continues until all of the remaining IF clauses are known to be true (in which case the hypothesis is true) or until no more rules apply (and the hypothesis is false).

A *data-driven expert system* uses a forward-chaining mechanism. The program begins with a list of facts that are known to be true. Each rule in the knowledge base is tested to see if all of its IF clauses are contained in the list of known facts. When such a rule is found, the system adds the THEN clauses from the rule to the list of known facts. All the rules in the knowledge base are scanned repetitively until no new facts can be concluded.

If the rules are formulated with weights or confidence factors to their conclusions, the system can produce an answer which is neither true nor false. Multiple answers are possible with degrees of importance or predictability.

The performance of an expert system may be increased by using heuristic rules to eliminate possible but unlikely solutions. The added knowledge is called meta-knowledge; in knowledge bases which use production rules, meta-knowledge is incorporated as meta-rules. These rules instruct the system how to choose which rule to use when more than one is relevant. For example, in an organic synthesis expert system a meta-rule might be IF multiple reactions have the same product THEN use the reaction with the highest yield.

11.6.5   Software and Hardware

Successful AI programs can be written in traditional computer languages but specialized AI languages are much more common in expert systems.

LISP (from LISt Processing) was developed in 1960 and is the most commonly used AI language in the United States [11.32]. It is used to define and manipulate irreducible objects associated with an alphanumeric label, called an atom. Atoms may be assigned values which are combinations of atoms and operators arranged in a list structure. Atoms are stored in memory and are located by means of pointers. Lists are represented as a collection of memory cells whose contents are pointers to other memory cells. Manipulation of lists involves manipulation of their pointers and application of some simple logic processes. LISP and LISP-like languages are typified by their interactive nature, the emphasis on symbolic expressions, and a tree-oriented approach to data structures.

PROLOG is used by most Europeans and Japanese. It is a higher level language than LISP, allowing a fact to be made in one statement where LISP would require many more programming steps. However LISP is more flexible and has more sophisticated programming aids. Advocates of PROLOG argue that it is more readable, provides a more suitable natural language interface, and can execute in a parallel fashion that improves speed of execution. It supports the easy construction and use of relational data bases.

PROLOG [11.12], [11.17], [11.43] is based on predicate calculus (see Section 11.6.2) which allows the user to program relationships and qualities or attributes.

Both compiler and interpreter forms (see Section 7.2) of LISP and PROLOG exist.

Expert systems can be run on traditional computer hardware but specialized minicomputers, usually LISP-oriented, give better results. PROLOG adapters for such systems are available. Microcomputer versions of both LISP and PROLOG are available and LISP compilers are offered for some general-purpose hardware.

The most common and cheapest method of developing a small expert system is by using a commercially available PC-based "shell". This is an empty knowledge base with its own representation language and inference engine. The user can concentrate on the development of the knowledge base, rather than the programming procedures. Shells have put the software in the hands of the expert rather than the knowledge engineer (see Section 11.6.3).

11.6.6   Advantages and Disadvantages

A well-written expert system performs consistently (although "mindlessly") for 24 h a day in an hostile environment. It can also release experts from tedious tasks for more important work. Many simultaneous users can be accommodated by an expert system.

Knowledge used by experts is generally learned through experience and is not written down. An expert system captures knowledge that could otherwise be lost. An average user may, however, be too trusting of the results output by the computer and may fail to ask for an explanation.

Characteristically, an expert system covers only a narrow domain. As the domain increases, the number of contingencies that can be covered decreases. Building and maintaining a large knowledge base is difficult. Even for a narrow domain, constructing the knowledge base is slow and laborious. Another problem is speed of execution.

Further problems and trends are discussed in [11.34].

11.6.7   Applications

Medical Diagnosis. The best known expert system in the field of medical diagnostics is MYCIN [11.9], [11.23] which advises on the treatment of bacterial infections, using backward-chaining techniques. It is written in LISP and has three modules: a consultation program (the user interface), an explanation program (the inference engine), and the knowledge acquisition program (the knowledge base and a maintenance interface). The physician enters data to record the patient's history and symptoms. If information is missing, MYCIN will request more data or infer it from the knowledge base. MYCIN then reports (and explains) its conclusions and suggests an appropriate treatment.

Librarianship and Information Retrieval. The online public access catalog (OPAC) should be amenable to the expert system approach [11.34], [11.40], [11.44]–[11.46]. An OPAC offers a user-friendly interface to locate documents in library catalogs. Visitors to the U.S. Library Corporation can use an intelligent CD-ROM-based

OPAC, the Bibliofile Intelligent Catalog, which uses both sound and graphics [11.47].

Online searching is also an area ripe for expert systems developments. The contents of a knowledge base for information retrieval from an online data base must be divided between system knowledge and subject knowledge, e.g., the command language and retrieval techniques as opposed to the choice of search terms and synonyms.

Some experimental information systems for online searching, are reviewed in [11.48], methodologies are discussed in [11.24]. The PLEXUS system [11.49], [11.50] links an intelligent search interface to a conventional data base, whereas the GENERIS system [11.51] incorporates the conventional structure of a relational data base but can also represent semantic relationships between items in the tables.

The TOME Searcher front end [11.24]–[11.26] allows people without experience to make use of data bases. It has semantic networks based on thesauri. The user can train the system to learn new terminology. The system runs on an IBM personal computer and has four components: a set of menu-driven functions which help the user set up search parameters; a natural language interface for search queries; a step-by-step question and answer module to refine the automatically generated search strategy; and a communications module to call up the search service, upload the search, and download the results. ESA, STN, DIALOG, and Maxwell Online can be accessed.

WILLETT and coworkers [11.52] have compared the knowledge-based approach to reference retrieval (exemplified by PLEXUS) with statistically based techniques (exemplified by Sheffield University's INSTRUCT program [11.53]–[11.55]).

INSTRUCT supports facilities for natural language query processing, best-match and cluster-based searching, user-initiated query expansion based on string similarity or term co-occurrence data, automatic relevance feedback based on probabilistic term weighting, and a browsing capability. An operational version for interactive browsing and ranking has been implemented in an industrial environment [11.56].

Ease-of-use and cognitive and behavioral aspects of retrieval are discussed in [11.57]–[11.60].

Recent publications have discussed the application of expert systems to library and information science work [11.15], [11.39], [11.40], [11.46], [11.61], [11.62].

Computer-Aided Synthesis Design. The first attempt to codify and organize organic syntheses was the Logic and Heuristics Applied to Synthetic Analysis (LHASA) program [11.63], [11,64]. The goal is a target molecule and the system works backwards through a synthesis tree by a method called retrosynthetic analysis, tracing possible precursors from a "transform library" of reactions. In Europe, LHASA UK handles the use of the program and maintenance of the transform library.

The Simulation and Evaluation of Chemical Syntheses (SECS) project [11.65]–[11.67] was initiated in 1969 to focus on stereochemistry, stereoelectronic effects, and other aspects not considered by LHASA. SECS also uses retrosynthetic analysis and a transform library. A consortium of seven major German and Swiss companies, the computer-assisted synthesis planning (CASP) project, has worked with SECS.

For other synthesis planning programs with different strategies see Section 10.9 [11.68]–[11.76].

Analytical Chemistry and Structure Elucidation. The interpretation of spectra has proved a fruitful area for the application of expert systems theory [11.32], [11.77]–[11.84]. The earliest venture was the DENDRAL (**dendr**itic **al**gorithm) project which used mass and carbon-13 NMR spectra [11.85]–[11.87].

BREMSER's work [11.88]–[11.91] is of interest because of the variety of spectral techniques, his substructure searching technology, and the commercial availability of the software (see Chap. 9). His IDIOTS (Infrared Spectra Documentation and Interpretation Operating with Transcripts and Structures) [11.88] system is, however, not commercially available.

CHEMICS (Combined Handling of Elucidation Methods for Interpretable Chemical Structures) uses IR, and carbon-13 and proton NMR spectra [11.92]–[11.94].

PAIRS (Program for the Analysis of Infrared Spectra) does not employ a data base of IR spectra but attempts to parallel the reasoning used in interpreting IR spectra [11.95]–[11.97].

McLAFFERTY and coworkers developed STIRS (Self-Training, Interpretive, and Retrieval System) to interpret unknown mass spectra [11.98].

CASE (Computer Assisted Structure Elucidation) [11.99] follows a new strategy [11.100]. It handles more than one type of NMR spectra and an IR spectrum interpreter is under development.

Other systems have been reported [11.101]–[11.104].

Other Applications. Other applications of expert systems are listed below:

MACSYMA, a symbolic manipulation system that functions as a mathematical aid.

R1, an expert system that aids in the configuration of Digital Equipment Corporation VAX Computer systems.

PROSPECTOR, an expert geologist system that seeks commercially exploitable mineral deposits.

MOLGEN, an expert system for designing experiments in molecular biology. The Imperial Cancer Relief Fund has a PROLOG system for protein topology [11.105].

The DOCENT expert system deals with macromolecules [11.106], [11.107]. It is a molecular modeling application which represents macromolecules by "generalized cylinders".

WIZARD is a symbolically based conformational analysis program which reasons about intra- and intermolecular forces [11.108]–[11.110].

The AIMB program (Analogy and Intelligence in Model Building) is a symbolic, nonnumerical approach to model building and conformational analysis [11.111], [11.112].

Metabolexpert predicts metabolic pathways by logic programming [11.113].

For other examples of systems in the molecular graphics area see [11.41], [11.114], [11.115]. There are also industrial applications of expert systems in chemistry [11.41], [11.115] and environmental systems [11.116].

11.7   Hypermedia [11.117]–[11.122]

The concept of an "intelligent data base" in a knowledge-based integrated information system, involves object orientation (see Section 7.6), expert systems, and hypermedia. The history of hypermedia goes back to 1945 and an article published by VANNEVAR BUSH [11.123]. BUSH had a vision of a device called Memex [11.123], [11.124], an electronic desk that could access the text of linking files in seconds. Inspired by BUSH's ideas TED NELSON coined the word "hypertext" in the 1960s, for a type of nonsequential reading and writing that links different text nodes [11.125]. Implementation of the theory was not possible until Office Workstations Ltd. International's Guide software and Apple Computer's HyperCard software were introduced in 1987.

A hypertext system allows the user to link pieces of information, creating trails through the associated materials. The idea is akin to the use of footnotes, references, and "See Also's" in a printed tool. The reader can "browse and hunt" in an electronic book. In early hypertext systems all links were preprogrammed. Interactive links in newer systems give the user more freedom.

A hypertext system may be envisaged as consisting of "cards" containing text or other information and connected by directed links. In most systems a card may have several outgoing links. Users navigate through the system by following the links. They may also backtrack by following the links they have used in the reverse direction.

"Landmarks" are especially prominent cards, for example, because they are directly accessible from many other cards. A hypertext system has two bidirectional navigational dimensions: a linear dimension used to move back and forward among the text pages of a given section; and a nonlinear dimension used for hypertext jumps.

Essential features of a hypertext system are an intuitive graphical user interface (mouse, buttons, icons, pull-down menus, consistency, Chap. 7), cross-hierarchical links, an object-oriented environment and scripting language, and a flexible format for information, not necessitating a formal data-base structure.

Hypermedia is an extension of the concept of hypertext and implies incorporation of color, graphics, images, sound, and animation (video).

Some problems still require resolution. Setting up a hypertext data base can be a very time-consuming proposition. The user may tend to feel lost while navigating the system. Text searching is limited and nontextual information (e.g., video) cannot be searched; a multiuser environment is not possible; there are high data storage requirements; and data storage and compression/decompression standards need establishing. The large storage space requirements for graphic and video have led to the association of hypertext systems with optical disc storage (see Section 6.3) and some CD-ROM data-base systems have hypertext-related retrieval software (see Chap. 8).

12   Basic Facts on Patents

12.1   Inventions and Innovations

Scientific research is directed towards enlarging the body of human knowledge, and its latest findings are published in scientific journals (Chap. 1). In contrast, industrial research aims at innovations, i.e., the introduction of new processes or products into technology or the improvement of old ones [12.1]–[12.4]. A rich source of innovative ideas is available in technical inventions disclosed in patent literature. Inventions constitute an important part of industrial property and are accorded legal protection on the basis of national laws and international patent conventions.

Although not every innovation emerges from an invention, and by far not all inventions mature into innovations, there is, as a corollary of progressive industrialization and ongoing competition, a strong tendency to seek legal protection for every promising technical invention emanating from industrial research. This tendency is increasing because, in view of the increasing efficiency of the methods of testing and analyzing materials, it has become more and more difficult to keep the composition of a product or its manufacturing process secret. Thus, patent literature is comprehensive with respect to inventions deemed exploitable in one way or another [12.5].

A national patent system comprises the whole range of legal and administrative aspects according to which patent applications are accepted and processed and rights are granted, enforced, or abolished, including the patent law and other pertinent laws, the rules of practice of the patent authorities, and court decisions on patent cases.

Correspondingly, the documents issued by the authorities administering intellectual property protection differ considerably, not only in language. A comprehensive treatment of the many existing patent systems cannot be given here. For details the reader is referred to the vast stock of special literature on industrial property protection. As an introduction, only the essential features of patent legislation and patent office procedures that are crucial for the understanding of patent literature and its role are outlined for some typical patent systems.

In a patent system, an invention can be patented only once. When different inventors have made the same invention and have applied for a patent it must be decided which inventor is entitled to recieve the patent. In most patent systems, the patent is granted to the inventor who was the first to file his application with the patent office. The *first-to-file principle* urges the inventor to hurry up with his preparations for filing his application, and in areas of high research activity there is a hectic race to ensure that inventions are patented ahead of competitors. Although there are public benefits from immediate filing and early publication of inventions, there is always a danger that the invention described is immature, its significance is not yet fully recognized, and the patent claims are not optimally worded for proper protection. Applicants will try to avoid these negative effects by drafting their patent applications in a vague and general style. This practice has impaired the reputation of patent literature as technical literature.

This problem is partly solved in those patent systems (e.g., Belgium, Germany, Japan, Sweden, United Kingdom) which offer the applicant a one year period during which he may improve his invention and the wording of his application and then refile it (*domestic priority*). No publication of the inventive idea made by himself or by any other person during the domestic priority period can be cited as prior art against the refiled application if the applicant has claimed domestic priority by referring to his older application. He has to prove that both the older and the younger applications are based on the same invention. Note that domestic priority (in contrast to Convention priority, Section 12.3) is related to refiling in the same patent system.

The patent law of the United States does not adhere to the first-to-file principle. If two or more applications claiming identical or similar inventions are received by the United States Patent and Trademark Office (USPTO), this authority will make an inquiry (*interference procedure*) as to who was the first to make the invention and to put it into practice (*first-to-invent principle*). The first inventor is granted the patent although he might not have been the first to file his application. In the United States, as long as a patent application has not been finally rejected by the examiner, it may be refiled (*continuation application*), even with certain amendments (new matter, *continuation-in-part application*).

12.2   Patents and Patent Systems

A patent is a temporary exclusive right (monopoly) granted to an inventor or applicant by the government in return for disclosing his invention to the public. During the term of a patent, only the patent owner is entitled to make use of the patented invention. A patent may, like other rights, be sold, bequeathed, or otherwise transferred to other parties. The patent owner may permit others to make use of the invention (patent license), usually on carefully stipulated terms.

To obtain protection for an invention, a formal application has to be filed with the relevant national or supranational (Section 12.3) patent office, together with a detailed description of the matter for which protection is claimed. In the course of the ensuing registration and granting procedure the patent office issues one or more documents which are referred to as patent literature. The information they contain is called patent information. These two terms are general terms covering not only patent specifications and published patent applications but also documents related to utility models, inventors' certificates, semiconductor topographies, and equivalent rights (Section 12.4).

Granting of patents has a long history, beginning in medieval England in 1331 [12.6] and Renaissance Venice in 1474 [12.7]. The first „modern" patent systems came into being in Britain (Statute of Monopolies) in 1624 [12.8], the United States (1790), and France (1791), when the fundamental principles of patenting that are still valid in our times were developed.

Patents are granted for technical inventions that are novel, non-obvious, and applicable in industry. Several categories of inventions (e.g., non-technical inventions) are excluded from patenting.

Novelty. An invention is deemed novel if it has not already been made known before the date the patent application is filed (or before the priority date) or, in technical jargon, if no *prior art* exists. Any divulgation, be it in written, oral, or other form, made by the inventor or anybody else before first filing constitutes prior art.

In some patent systems novelty is defined slightly differently. A short preclusive period is granted to the inventor with respect to his own publication made before filing.

Even an application filed earlier by another person but not yet published by the patent office can be cited as prior art against the younger application if it claims the same invention. Novelty is assumed to exist if no prior art has been detected by the patent examiner (if there is substantive examination). If prior art comes to the knowledge of the patent office after grant, the patent might be restricted or even declared invalid. Prior art may be discovered in any technical literature. Patent literature is the most promising source material to be searched for prior art.

Non-obviousness. Non-obviousness of an invention means that its crucial idea would not have been immediately apparent to persons skilled in the art, but embodies an inventive step. To determine whether an invention is obvious or not, it is judged from the point of view of a specialist whose technical background is the field the invention belongs to (patent examiners are specialists). The technical background is based on the pertinent technical literature (especially the patent literature) and professional experience.

Intellectual property protection is part of a country's judicial system which in turn is deeply influenced by the country's history, traditions, and political situation. Therefore, laws pertaining to intellectual property vary from country to country, and many different patent systems exist [12.9]–[12.11].

Registration Patents. In a number of patent systems, patents are granted without any substantive examination of inventions as to novelty, obviousness, or applicability; granting is simply a registration procedure. Argentina, Belgium, France, Greece, Italy, Luxembourg, Spain, and South Africa, for example, are countries that grant registration patents. Their granting procedures differ in several details, e.g., in France an application may be rejected on the basis of the search report. Assessment of a patent as to its enforceability is made in court in an infringement suit. Switzerland has a two-tier patent system: substantive examination is required only for applications pertaining to time measurement or non-mechanical textile finishing. In every other area of technology Swiss patents are registration patents.

Patent Examination. In other patent systems, substantive examination of the invention described in a patent application as to *novelty, obviousness, industrial applicability*, and other requirements for patentability is performed by the patent office before a patent is granted. In general, two approaches may be distinguished:

1) Examination is carried out automatically for every application received. When an invention is deemed patentable, a patent is granted on it and a patent specification is issued (e.g., in Austria, the Soviet Union, and the United States).
2) Examination begins only after it has been requested (*deferred examination*), a petition for examination has to be made within a certain period of time (e.g., in Germany, the Netherlands, Japan 7 years, in the People's Republic of China 3 years from the filing or priority date, in the United Kingdom 6 months from the publication date of the search report). If the applicant fails to request examination the application is deemed abandoned.

In patent systems with deferred examination, the unexamined patent application, together with a title page (Section 16.2) provided by the patent office, is published 18 months from the filing or priority date (e.g., in the People's Republic of China, France, Germany, Canada, Japan, the Netherlands, Sweden, the United Kingdom; Switzerland in the areas of time measurement and nonmechanical textile finishing only). For a considerable number of these unexamined applications the publication deadline is not met by the patent offices.

The published, unexamined patent applications are the earliest descriptions of new inventions available to the public. Therefore, they are of utmost importance to patent information management (Chap. 13).

In some patent systems, the search for prior art and examination is performed in two separate steps. Certain patent offices (e.g., the German Patent Office) offer applicants and any other interested parties searches in their files for prior art with respect to patent applications filed before examination has been requested. Other patent offices (e.g., the British Patent Office) carry out prior art searches automatically when applications are received. In both cases, the search reports are announced in the patent gazettes and are made available to the public.

In some patent systems, e.g., in the People's Republic of China, Germany, Japan, Austria, the United Kingdom and Switzerland (for examined cases only), the public is invited to notify the patent office within a certain period of time (*opposition period*) of any objections to the patentability of the invention. Even after the opposition period has expired, a patent can be attacked in a *nullity suit* on the grounds of non-patentability, which is also usually based on prior art retrieved from the patent literature.

In the United States a patent may undergo *reexamination*. When it is upheld, a new patent specification (reexamination certificate) with revised wording is published.

Protection of the invention begins with the granting of the patent and has a retrospective effect from the date of filing. Patents have a maximum term: 20 years from the application date in several countries, 17 years from the patenting date in the United States, different terms in other countries. In certain areas of technology (e.g., pharmaceuticals) the period between filing the patent application and marketing the corresponding product (the lead time) is excessively long because of the lengthy testing procedures and the strict regulations of the national health authorities. In such cases, an extension of the patent term may be applied for in the United States, France, and Japan. A motion has recently been directed to the European Commission for adopting the same practice [12.12].

In most patent systems, patent owners must pay annual fees for keeping their patents in force. If they fail to do so their patents will expire. When a patent has expired anybody may make use of the invention. In socialist countries, an inventor employed in a government-owned enterprise is not entitled to apply for a patent but he is awarded an inventor's certificate. In this way, the government obtains a license for making use of the invention in its factories and, in return, undertakes to exploit diligently the invention and pay the inventor adequate compensation. As the patent laws of some socialist countries are now being adapted to western principles (e.g., in the Soviet Union) or have already been changed, inventors' certificates will probably disappear.

The scope of protection of a patent is governed by the patent claim(s) as allowed by the patent office. A patent owner may take any person or organization that infringes his patent to court.

12.3   International Patent Conventions

A patent granted by a national authority is, of course, enforceable only on the territory of that particular country. If an applicant also wants to obtain protection for his invention in other countries, he has to file patent applications with the respective patent offices as well. Multinational patent filing has been a difficult, expensive, and risky procedure. To overcome this, a series of international patent conventions have been concluded by national governments. Only those most relevant to patent information management are specified here.

12.3.1   Paris Convention

In 1883 a very important agreement, the *Paris Convention for the Protection of Industrial Property*, came into being. It has since undergone a number of revisions, and governments of many states have successively signed it to form the Paris Convention Union. In January 1991, 100 countries had accorded to it.

From the point of view of patent information management, the most significant achievement of the Paris Convention is the legal instrument of *Convention priority*: An applicant who has filed a patent application in one of the member states of the

Paris Union is granted a one year priority period, reckoned from the date of first filing, in which he may file applications based on the same invention in other member states of the Paris Union without his first application or any other description of that invention made public during the priority period being considered as an anticipation of his follow-up applications.

Convention priority is granted only when it is explicitly claimed by the applicant. The applicant has to prove that the invention described in the follow-up applications is identical with that on which the first-filed application was based. The Convention priority takes effect only for follow-up applications in different countries, whereas the domestic priority (Section 12.2) pertains to follow-up applications in the same country.

The applicant may combine more than one first-filed application into one follow-up application which then claims more than one priority. The priority period of one year must not, however, be exceeded; domestic priority and convention priority cannot be combined into a period of more than one year.

The Paris Convention is supervised and developed by the Union's International Bureau run by the *World Intellectual Property Organization* (WIPO), Geneva. The WIPO was established by an international agreement concluded in 1967 and has also been accorded several other responsibilities [12.13], [12.14] (see below). The WIPO Agreement has been signed by 125 states (January 1991).

12.3.2   Patent Cooperation Treaty

The Patent Cooperation Treaty (PCT) of 1970 forms the basis of a supranational patent system. In the PCT system, an applicant can apply for a patent in any of the national patent systems of the PCT member states (and the European patent system) he may choose (designated states) by filing just one application.

Depending on his country of residence, the applicant has to file his application in a specified language (English, German, French, Spanish, Russian, or Japanese) and to direct it to the Accepting Office specified in the PCT rules. From there, the application is forwarded to the WIPO which allocates an international application number to it and arranges for an international search report. Eighteen months from the filing date or, if claimed, the priority date, the application and the search report are published and the publication is announced in the *PCT Gazette*. The WIPO may also order an international examination report to be prepared by the International Examination Authority specified in the PCT rules. The PCT application then leaves the international phase and enters the national phase, i.e., it is processed by the patent offices of the designated states as a national patent application.

Presently (January, 1991), 45 countries are PCT members. Although the PCT procedures are rather complicated, the number of PCT applications filed annually has, after a slow start, reached remarkable levels (19 159 applications in 1990 [12.15]). In the PCT system, only the application procedures are centralized. The competences of examination and grant still lie with the national patent authorities.

12.3.3   European Patent Convention

The European Patent System, based on the European Patent Convention of 1973, goes a step further. European patent applications can be filed with the European Patent Office (EPO), either with its headquarters at Munich (Germany), its Directorate General 1 (DG 1) at The Hague (Netherlands), or with some other national authorities in member states of the European system which pass the applications on to the EPO.

Presently (January, 1991), 14 European states adhere to the European Patent system. When applying for a European patent, an applicant must designate the national patent systems in which he wants to have his invention protected. The official languages of the European Patent System are German, English, and French [12.16].

When an application has been accepted by the EPO, a search for prior art is carried out in the files of DG 1, resulting in a search report. Eighteen months from the filing date or, if priority is claimed, from the priority date, the application, a title page, and the search report are published and the publication is announced in the European patent gazette.

Often, DG 1 has been unable to finish the search report on time; the report is then published separately. Within six months from the publication of the search report, examination must be applied for, otherwise the application is deemed abandoned.

When the invention meets the requirements for patentability, a European patent is granted on it and a European patent specification is issued in the language in which the application has been filed and with the patent claim(s) in German, English and French. The grant is announced in the European patent gazette (*European Patent Bulletin*). Within nine months from the date of grant, an opposition may be filed with the EPO. If the claims are restricted by the opposition procedure, a revised European patent specification is issued.

The opposition procedure is handled by the EPO, but once it has been granted, a European patent „disintegrates" into a set of national patents enforceable in the member states the applicant has designated. Thus, the member states have delegated the application and granting procedures to the EPO. Nullity suits and patent litigations remain in the national realm.

The European Patent System has been amazingly well received by the applicants' community: In 1990, 62 788 patent applications were filed with and 24 757 patents were granted by the EPO [12.17].

Provided a country is a member of the PCT, and the European systems, national patents can originate by three different routes: the national, the PCT, and the European routes. Therefore, the PCT and the European applications as well as the European patent specifications must be watched as closely as the national patent documents.

12.3.4   The Common Market Patent Convention

The Convention for the European Patent for the Common Market of 1975 has been signed and ratified by nine states. As not all EC member states have acceded yet, this Convention has not yet come into force. Its object is to create a truly international patent enforceable in all EC countries and based upon a unified, independent, and fully developed judicial system. There is a good chance that the EC Patent Convention will soon be realized in stages, starting with those countries that are willing to join and leaving the door open for those that are presently not in a position to accede.

12.3.5   Other Patent Conventions

In addition to the supranational patent systems mentioned in Sections 12.3.2–12.3.4, there are other patent cooperation schemes for two groups of African states, ARIPO (1976/1982) and OAPI (1979), and for the socialist countries forming the COMECON bloc. As the latter has recently been dissolved, the future of this patent convention (*Havana Convention*) is not clear.

The Convention on the Unification of Certain Points of Substantive Law on Patents (Strasbourg Convention) of 1963, presently signed by eleven European states, has paved the way for a considerable harmonization of the patent laws of its signatory states and has been an important step towards the European patent system.

The International Convention on the Protection of Topologies of Integrated Circuits is dealt with in Section 12.4.

12.4   Petty Patents

Besides their patent systems, some countries have, for many years, been granting „smaller" technical intellectual property rights (i.e., monopolies with shorter terms or for special categories of inventions) in patent jargon known as petty patents. Whereas the recent developments of the full-fledged patent systems have been geared towards harmonization and adaptation, petty patents have existed in their own right, mostly tailored to the specific requirements of those countries' industrial environments and traditions. Therefore, petty patents are even more diverse in character than „normal" patents and defy systematization.

Nevertheless, petty patents have become increasingly important because in certain countries they have been highly valued by the legislative bodies and vested with additional power, in effect as a counterweight for the increasing internationalization of the patent systems.

The most significant group of petty patents are the *utility models* which exist in Germany, France, Italy, Poland, Portugal, Sweden, Spain, Japan, the People's Republic of China, the Philippines, South Korea, Taiwan, Brazil, Uruguay, Morocco, and Australia. No comprehensive treatment can be given here, interested readers should consult the pertinent specialized literature. The features of the German and the Japanese utility models are explained briefly as typical examples.

German Gebrauchsmuster. The German Gebrauchsmuster (German utility model) was created in 1891 and, until recently, underwent few changes. As its name suggests, it was meant as a low-cost property right covering short-lived technical inventions related to new configurations of tools and articles of daily use. The German utility model law was revised in 1987 and 1990. Now, Gebrauchsmuster are granted on any technical inventions (except processes) and their term has been extended to ten years at most. Gebrauchsmuster are registry patents. Registration takes place within a few weeks from filing, providing immediate protection. The Gebrauchsmuster descriptions are not published, but photocopies of the applications are available.

Japanese Jitsuyoushinan. The Japanese utility model, jitsuyoushinan, is similar to a „normal" patent. The application is subjected to substantive examination, its maximum term is 10 years from publication of the application or 15 years from its filing date. Both the application and the specification of the the the granted right are officially published.

Utility model descriptions are a neglected species of patent literature. Despite the fact that their importance is increasing and they are acknowledged as official documents and constitute a considerable proportion of the patent literature, no major commercial information service (Chap. 20) has been interested in them. Also, most patent searchers simply ignore utility models.

In contrast to utility models, *integrated circuits* have only recently become a topic of intellectual property. In 1989, an *International Convention on the Protection of Topologies of Integrated Circuits* was drawn up in Washington, DC, United States. The Convention pertains to the legal protection of three-dimensional microelectronic semiconductor devices for a term of at least 8 years, covering their manufacture, import, and sale. A considerable number of countries have acceded, e.g., Austria, Belgium, Canada, Denmark, France, Germany, the United Kingdom, Italy, Japan, the Netherlands, the People's Republic of China, Spain, Sweden, Switzerland, and the United States. In most of these countries, the topologies disclosed in applications filed with the patent offices are registred and made available to public inspection. Generally, the term of protection is 10 years.

12.5   The Significance of Patents to Industry

By safeguarding advantageous positions gained by creative persons or enterprises in economic life and setting the rules for obtaining and defending these positions, legal protection of technical intellectual property has played a significant role in shaping contemporary industrialized societies. The importance of intellectual property protection to industry is enhanced by the increasing complexity of international trade relations, the growing size of markets, and diminishing difficulties of gaining entry to them. As a corollary, investments in research and product development have soared, intensifying the consequences of risky decisions in research and product strategies. Therefore, it is of vital importance for industry to be offered the chance of preventing competitors, who have not taken any risks and not incurred research and development expenses, from exploiting other companies' inventive products or processes for a limited amount of time.

Patent protection is particularly important in the chemical industry because many chemical products and processes are long-lived, requiring expensive specialized equipment and long-term development with long pay-out times. This is one of the reasons why many chemical patents are kept in force by their owners for their entire terms. Strong and long-kept patents are, of course, also a welcome source of license revenues. Patent and licensing policies play an instrumental role in marketing and technology transfer as well as in creating and maintaining employment [12.18]–[12.21]. The descriptions of competitors' inventions disclosed in the patent literature provide a strong stimulus to the creativity of research and development staff of an industrial company.

For all these reasons, the chemical industry has always placed a great deal of emphasis on patents. Patent information management keeps executives, researchers, patent specialists, and other employees in the company informed on patent literature. Patent information management is at the heart of industrial information management [12.22].

13 The Functions of Patent Literature

Patent literature has two inherently different functions: the *information function* and the *legal function* that are closely related to one another. In order to be patentable, an invention must be disclosed fully and clearly so that it is workable by a person skilled in the art. Non-workability is a sufficient reason for the rejection of a patent application or the revocation of a patent. Nothing is protected that is not disclosed. Each of the two functions poses specific problems for patent information management.

13.1 The Information Function

Understandably, industrial enterprises are reluctant to reveal any technologically relevant results of their research and development activities. If this is done, the medium of divulgation is the patent literature because there is no other way of obtaining patents. In the information network of technology the patent literature thus assumes the role that the scientific literature plays in the information system of science.

The disclosures forced from the applicant by the patent systems are, of course, limited to information that is essential for obtaining patents. As descriptions of inventions the disclosures are addressed to persons skilled in the art and are written in expert jargon, rather than in plain language easily understood by a layman. Consequently, the information function appeals primarily to researchers and developers [13.1], [13.2] forming a special user community to be served by the patent information specialists. As with the scientific literature, the information needs of the users are satisfied in two ways: by current awareness and by retrospective searches.

Current Awareness. Current awareness is related to the currently issued patent literature which has to be made known to all those who it concerns, in a timely, reliable, and comprehensive manner. The ways and means of achieving this are explained in Chapter 19. Some aspects of providing information to the research and development user community should, nevertheless, be mentioned here.

Patent literature is a highly specialized brand of technical literature and can only be reliably channeled to its target groups if the personnel handling it are sufficiently trained in the pertinent technologies so that they can understand its contents and realize its pertinence to the individual readers and their subjects of research. In addition to patent documents, the scientists should be encouraged to read abstracts bulletins (Section 19.3.2) that cover wide, interdisciplinary areas of technology with some bearing on their work. Patent literature, in contrast to other technical literature, has a very low redundance rate. As a consequence of the requirement of novelty, an invention can be patented in a patent system only once; it is, therefore, highly unlikely that a technical detail disclosed in a patent document will also be found in other patent documents. It is for this reason that patent literature relevant to a reader's task should be delivered to him completely. It should be left to the reader to make the final selection. Redundance might occasionally occur because patent applications are issued unexamined in some patent systems. Redundance is unavoidably caused by follow-up applications in different patent systems for the same invention, but this is not a problem in retrospective searches as will be explained in Chapter 19.

Retrospective Search. There is little chance of meeting technical information disclosed in the patent literature in the journal literature [13.3]–[13.5]. It is disclosed much earlier in the patent literature than elsewhere, and in many cases it is never disclosed in any other media. In contrast to the current awareness activities triggered by incoming documents, retrospective searches are problem-oriented [13.6]. They have to help a research scientist starting work in an unfamiliar area to tackle the problem at hand, to obtain a sound concept for its solution, and to work out a description of his invention on which a patent application can be based.

When embarking on research in a new field, a scientist is well advised to ask for a comprehensive search that reveals all the prior art. A careful study of patent literature dating back a number of years is also worthwhile. Many ideas disclosed in documents for which the corresponding patents have already expired might well have been ahead of their time and might not have been put into practice for technical or economic reasons. Circumstances might have changed in the meantime, and reconsideration of these ideas might be worthwhile. Old patent literature is not covered by most commercial online databases; it is only searchable manually in patent document collections or card files kept in industrial information departments.

When an invention assumes concrete form, additional searches on specific details might be necessary. When the text of a patent application is being formulated, a further search based on the drafted patent claim(s) should be carried out in anticipation of the patent examiner's objections.

13.2   The Legal Function

The legal function is the aspect of the patent literature that is particularly relevant to the tasks of the patent specialists [13.7] working in the patent departments of industrial companies. They are concerned with filing applications based on inventions made in their enterprises, obtaining patents, legally safeguarding products to be launched and processes to be put into operation, defending the patents obtained against infringement, or licensing them to others. They also have to watch the patent activities of competitors, file oppositions and nullity suits against patents regarded as granted wrongly, ward off attacks on patents of the company, and negotiate terms for external licenses.

In all these activities patent literature plays a vital role and must therefore be supplied to the patent specialists accurately and timely by patent information management. Apart from current awareness, retrospective searches concentrate on bibliographic patent data (Section 19.2) or technical disclosure (Section 19.3) for retrieving prior art or existing patents of others that could impede the company's activities. The measures to be taken are explained in Chapter 19.

13.3   Origin and Significance of Patent Literature in Patent Procedures

The publication of patent documents is intimately linked to the successive steps leading from filing a patent application to obtaining an enforceable patent. As already mentioned in Section 12.2, in patent systems with deferred examination a patent application is published 18 months from its filing date or its (earliest) priority date if domestic (Section 12.2) or Convention priority (Section 12.3) is claimed. In addition to these patent systems, others publish patent documents fairly early, e.g., Ireland, Portugal, or Australia.

The publication of a patent application or patent specification is announced in the patent system's gazette (Section 18.1). Figure 7 shows the title page of a published European patent application, and Figure 8 the corresponding entry in the *European Patent Bulletin*.

The publication of a patent application does not constitute a monopoly, it is simply an early public warning that a monopoly has been applied for and might be granted. Published patent applications are the most important documents in patent information management because they are the earliest disclosures of new inventions (Section 13.1). Since the inventions they describe have not yet been examined for patentability, they constitute the broadest presentations of inventions, i.e., they are the documents which carry the highest information contents in patent literature: When an application has been filed no additional facts must be added to the description, and the applicant is not allowed to broaden the claim(s) beyond the scope of the original application. During the examination process, however, the scope of the claim(s) might well be restricted on the grounds of prior art, and sometimes the description of the invention also has to be adapted accordingly. Consequently, published patent applications play a major role in current awareness, in patent

(19) Europäisches Patentamt
European Patent Office
Office européen des brevets

(11) Publication number: **0 327 356**
A1

(12) **EUROPEAN PATENT APPLICATION**

(21) Application number: **89301001.7**

(22) Date of filing: **02.02.89**

(51) Int. Cl.⁴: **C 07 D 301/10**
B 01 J 23/66, B 01 J 23/50

---

(30) Priority: **03.02.88 CN 88100400**

(43) Date of publication of application:
09.08.89 Bulletin 89/32

(84) Designated Contracting States: **DE GB NL**

(71) Applicant: **CHINA PETROCHEMICAL CORPORATION**
24 Xiaoguan Street Anwai
Beijing (CN)

(84) Designated Contracting States: **DE GB NL**

(71) Applicant: **RESEARCH INSTITUTE OF BEIJING
YANSHAN PETROCHEMICAL CORPORATION**
9 Fonghuanting Road
Yanshan District Beijing (CN)

(84) Designated Contracting States: **DE**

(72) Inventor: **Jin, Jiquan** Research Institute of Beijing
Yanshan
Petrochemical Corporation
Yanshan District Beijing (CN)

Jin, Guoquan Research Institute of Beijing Yanshan
Petrochemical Corporation
Yanshan District Beijing (CN)

Xu, Yong Research Institute of Beijing Yanshan
Petrochemical Corporation
Yanshan District Beijing (CN)

Shang, Liandi Tianjin Chemical Research Institute
of Ministry of China Chemical Industry
Tianjin (CN)

Luo, Guochun Tianjin Chemical Research Institute
of Ministry of China Chemical Industry
Tianjin (CN)

(74) Representative: **Knowles, Audrey Elizabeth**
624 Pershore Road
Selly Park Birmingham B29 7HG (GB)

---

(54) **High efficiency silver catalysts for the production of ethylene oxide via ethylene oxidation.**

(57) This invention relates to a process of preparing silver-containing catalysts and their carriers for the production of ethylene oxide via ethylene oxidation and also to the applications of said catalysts in producing ethylene oxide . A commercial trihydrated α-alumina , boehmite , carbonaceous materials , a fluxing agent , fluoride and a binder are mixed with water , kneaded andextruded to form strips which are cut and shaped. The shaped bodies are then dried , calcined and converted to α-alumina bodies i.e.carriers. This process is characterized by using trihydrated α-alumina , boehmite alumina and carbonaceous materials which have a good matching of particle sizes and proportions in preparing alumina carriers with the following pore structure:

specific surface area	0.2 - 2 m²/g
pore volume	> 0.5 ml/g
pore radius	> 30 μ, 25-10% of total volume
	< 30 μ, 75-90% of total volume

Said alumina carriers are impregnated with silver compounds and promoters, and then dried , activated and used in ethylene oxidation for making ethylene oxide. The selectivity of the catalyst reaches from 83 to 84 percent.

Figure 7. The title page of a European patent application

Europäisches Patentamt

⑩   European Patent Office   ⑪ Publication number: **0 097 504**
Office européen des brevets   **B1**

⑫   **EUROPEAN PATENT SPECIFICATION**

⑤ Date of publication of patent specification: **07.01.87**   ⑤ Int. Cl.⁴: **B 60 R 19/18**

㉑ Application number: **83303515.7**

㉒ Date of filing: **17.06.83**

㊴ Core material for automobile bumpers.

㉚ Priority: **19.06.82 JP 105660/82**

㊽ Date of publication of application:
04.01.84 Bulletin 84/01

㊺ Publication of the grant of the patent:
07.01.87 Bulletin 87/02

㊾ Designated Contracting States:
DE GB IT

㊿ References cited:
DE-A-2 751 077
DE-B-1 794 025
Chemical Abstracts vol. 71, no. 13, 29
September 1969, Columbus, Ohio, USA; H.H.
Lubitz "Minicel polypropylene foam ", page 26,
column 2, abstract no. 62035q
Chemical Abstracts vol. 82, no. 10, 10 March
1975, Columbus, Ohio, USA; D.P. HUG et al.
"Polyolefin structural foam for automotive
use", page 86, column 1, abstract no. 59198s
Patent Abstracts of Japan, vol. 7, no. 97 , 23
April 1983, page 115C163
Patent Abstract of Japan, vol. 7, no. 100, 28
April 1983, page 9C164

㉲ Proprietor: **Japan Styrene Paper Corporation**
1-1, 2-chome, Uchisaiwai-cho
Chiyoda-ku Tokyo (JP)
㉲ Proprietor: **NISSAN MOTOR CO., LTD.**
No.2, Takara-cho, Kanagawa-ku
Yokohama-shi Kanagawa-ken 221 (JP)

㉲ Inventor: **Adachi, Akira**
515-14, Joza
Sakura-shi Chiba-ken (JP)
Inventor: **Kubota, Takashi**
3268-249, Nishinomiya-cho
Utsunomiya-shi Tochigi-ken (JP)
Inventor: **Okada, Yukio**
Shiroyama-Danchi 2-104 4589, Ohba
Fujisawa-shi Kanagawa-ken (JP)
Inventor: **Miyazaki, Kenichi**
4-14-11, Naka-machi
Machida-shi Tokyo (JP)
Inventor: **Hagiwara, Taro**
7-36-1-518, Sagamiohno
Sagamihara-shi Kanagawa-ken (JP)

㉴ Representative: **Myerscough, Philip Boyd et al**
J.A.Kemp & Co. 14, South Square Gray's Inn
London, WC1R 5EU (GB)

Note: Within nine months from the publication of the mention of the grant of the European patent, any person may give notice to the European Patent Office of opposition to the European patent granted. Notice of opposition shall be filed in a written reasoned statement. It shall not be deemed to have been filed until the opposition fee has been paid. (Art. 99(1) European patent convention).

Figure 9. Title page of a European patent specification

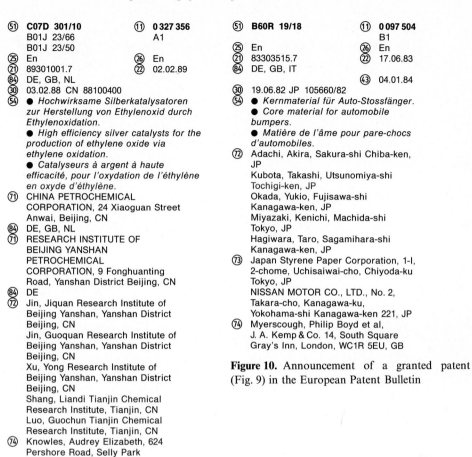

(51) **C07D 301/10**
B01J 23/66
B01J 23/50
(25) En
(21) 89301001.7
(84) DE, GB, NL
(30) 03.02.88 CN 88100400
(54) ● *Hochwirksame Silberkatalysatoren zur Herstellung von Ethylenoxid durch Ethylenoxidation.*
● *High efficiency silver catalysts for the production of ethylene oxide via ethylene oxidation.*
● *Catalyseurs à argent à haute efficacité, pour l'oxydation de l'éthylène en oxyde d'éthylène.*
(71) CHINA PETROCHEMICAL CORPORATION, 24 Xiaoguan Street Anwai, Beijing, CN
(84) DE, GB, NL
(71) RESEARCH INSTITUTE OF BEIJING YANSHAN PETROCHEMICAL CORPORATION, 9 Fonghuanting Road, Yanshan District Beijing, CN
(84) DE
(72) Jin, Jiquan Research Institute of Beijing Yanshan, Yanshan District Beijing, CN
Jin, Guoquan Research Institute of Beijing Yanshan, Yanshan District Beijing, CN
Xu, Yong Research Institute of Beijing Yanshan, Yanshan District Beijing, CN
Shang, Liandi Tianjin Chemical Research Institute, Tianjin, CN
Luo, Guochun Tianjin Chemical Research Institute, Tianjin, CN
(74) Knowles, Audrey Elizabeth, 624 Pershore Road, Selly Park Birmingham B29 7HG, GB

(11) **0 327 356**
A1
(26) En
(22) 02.02.89

Figure 8. Announcement of patent application (Fig. 7) in the European Patent Bulletin

(51) **B60R 19/18**
(25) En
(21) 83303515.7
(84) DE, GB, IT
(30) 19.06.82 JP 105660/82
(54) ● *Kernmaterial für Auto-Stossfänger.*
● *Core material for automobile bumpers.*
● *Matière de l'âme pour pare-chocs d'automobiles.*
(72) Adachi, Akira, Sakura-shi Chiba-ken, JP
Kubota, Takashi, Utsunomiya-shi Tochigi-ken, JP
Okada, Yukio, Fujisawa-shi Kanagawa-ken, JP
Miyazaki, Kenichi, Machida-shi Tokyo, JP
Hagiwara, Taro, Sagamihara-shi Kanagawa-ken, JP
(73) Japan Styrene Paper Corporation, 1-I, 2-chome, Uchisaiwai-cho, Chiyoda-ku Tokyo, JP
NISSAN MOTOR CO., LTD., No. 2, Takara-cho, Kanagawa-ku, Yokohama-shi Kanagawa-ken 221, JP
(74) Myerscough, Philip Boyd et al, J. A. Kemp & Co. 14, South Square Gray's Inn, London, WC1R 5EU, GB

(11) **0 097 504**
B1
(26) En
(22) 17.06.83
(43) 04.01.84

Figure 10. Announcement of a granted patent (Fig. 9) in the European Patent Bulletin

examiners' files, and as a source of information retrieval. When, after thorough examination, an invention meets the requirements of patentability, a patent is granted. A patent specification with the claim(s) approved by the examiner is issued (Figure 9 shows the title page of a European patent specification) and the grant is announced in the patent gazette (Fig. 10). In patent systems offering the opportunity of opposition (Section 12.2), the opposition period starts with the grant. For opposition, the most productive source of prior art not yet considered is the patent literature. Presentation of relevant prior art by an opponent might lead to partial or total revocation of the patent. When the patent is partially revoked, a revised patent specification is issued. Figure 11 displays the title page of a revised European patent specification.

(19) Europäisches Patentamt
European Patent Office
Office européen des brevets

(11) Veröffentlichungsnummer: **0 160 783**
B2

(12) **NEUE EUROPÄISCHE PATENTSCHRIFT**

(45) Veröffentlichungstag der neuen Patentschrift: **02.01.91**

(51) Int. Cl.^{5}: **B 41 M 5/00,** B 01 J 13/00

(21) Anmeldenummer: **85101357.3**

(22) Anmeldetag: **08.02.85**

(54) **Verfahren zum Einkapseln gelöster Reaktionspartner von Farbreaktionssystemen, die danach erhältlichen Kapseln sowie deren Verwendung in Farbreaktionspapieren.**

(30) Priorität: **09.03.84 DE 3408745**
19.11.84 DE 3442268

(43) Veröffentlichungstag der Anmeldung:
13.11.85 Patentblatt 85/46

(45) Bekanntmachung des Hinweises auf die Patenterteilung:
15.06.88 Patenblatt 88/24

(45) Bekanntmachung des Hinweises auf die Entscheidung über den Einspruch:
02.01.91 Patentblatt 91/01

(84) Benannte Vertragsstaaten:
AT BE CH DE FR GB IT LI LU NL SE

(56) Entgegenhaltungen:
DE-A-2 447 103
DE-A-3 346 601
DE-B-2 306 454
DE-C-2 134 236
FR-A-2 083 121
FR-A-2 235 804
US-A-3 578 844
US-A-3 940 275
US-A-4 333 849

Römpps Chemie-Lexikon, 8. Auflage, 1981, S. 1323-1326

(73) Patentinhaber: **Papierfabrik August Koehler AG**
Postfach 1245
D-7602 Oberkirch (DE)

(72) Erfinder: **Pietsch, Günther**
Burgwedeler Strasse 150
D-3004 Isernhagen HB (DE)
Erfinder: **Hartmann, Claus**
Kirchhöffner Strasse 6
D-3000 Hannover 91 (DE)

(74) Vertreter: **Hagemann, Heinrich, Dr. et al**
Patentanwälte Hagemann & Kehl Postfach 860329
D-8000 München 86 (DE)

EP 0 160 783 B2

Courier Press, Leamington Spa, England.

Figure 11. Title page of a revised European patent specification

14   The Volume of Patent Literature

Worldwide, intellectual property titles are issued by some 180 national and 4 supranational authorities (Table 11). They presently produce around $\times 10^6$ patent documents annually. The total amount of patent literature published so far is estimated at 25×10^6 documents, representing an invaluable stock of technical knowledge.

Table 11. List of countries and other entities issuing or registering industrial property titles, in the order of their codes *

Code	Country	Code	Country	Code	Country
AE	United Arab Emirates	CA	Canada	ES	Spain
AF	Afghanistan	CF	Central African	ET	Ethiopia
AG	Antigua and Barbuda		Republic		
AI	Anguilla	CG	Congo	FI	Finland
AL	Albania	CH	Switzerland	FJ	Fiji
AN	Netherlands Antilles	CI	Côte d'Ivoire	FK	Falkland Islands
AO	Angola	CL	Chile		(Malvinas)
AR	Argentina	CM	Cameroon	FR	France
AT	Austria	CN	China		
AU	Australia	CO	Colombia		
		CR	Costa Rica	GA	Gabon
		CS	Czechoslovakia	GB	United Kingdom
BB	Barbados	CU	Cuba	GD	Grenada
BD	Bangladesh	CV	Cape Verde	GH	Ghana
BE	Belgium	CY	Cyprus	GI	Gibraltar
BF	Burkina Faso			GM	Gambia
BG	Bulgaria			GN	Guinea
BH	Bahrain	DD	German Democratic	GQ	Equatorial Guinea
BI	Burundi		Republic	GR	Greece
BJ	Benin	DE	Germany, Federal	GT	Guatemala
BM	Bermuda		Republic of	GW	Guinea-Bissau
BN	Brunei Darussalam	DJ	Djibouti	GY	Guyana
BO	Bolivia	DK	Denmark		
BR	Brazil	DM	Dominica	HK	Hong Kong
BS	Bahamas	DO	Dominican Republic	HN	Honduras
BT	Bhutan	DZ	Algeria	HT	Haiti
BU	Burma			HU	Hungary
BW	Botswana	EC	Ecuador		
BZ	Belize	EG	Egypt		

Table 11. (Continued)

Code	Country	Code	Country	Code	Country
ID	Indonesia	MT	Malta	SR	Suriname
IE	Ireland	MU	Mauritius	ST	Sao Tome and Principe
IL	Israel	MV	Maldives	SU	Soviet Union
IN	India	MW	Malawi	SV	El Salvador
IQ	Iraq	MX	Mexico	SY	Syria
IR	Iran (Islamic Republic of)	MY	Malaysia	SZ	Swaziland
		MZ	Mozambique		
IS	Iceland			TD	Chad
IT	Italy	NE	Niger	TG	Togo
		NG	Nigeria	TH	Thailand
JM	Jamaica	NI	Nicaragua	TN	Tunisia
JO	Jordan	NL	Netherlands	TO	Tonga
JP	Japan	NO	Norway	TR	Turkey
		NP	Nepal	TT	Trinidad and Tobago
		NR	Nauru	TV	Tuvalu
KE	Kenya	NZ	New Zealand	TW	Taiwan, Province of China
KH	Democratic Kampuchea				
KI	Kiribati	OM	Oman	TZ	United Republic of Tanzania
KM	Comoros				
KN	Saint Christopher and Nevis	PA	Panama		
KP	Democratic People's Republic of Korea	PE	Peru	UG	Uganda
		PG	Papua New Guinea	US	United States of America
KR	Republic of Korea	PH	Philippines		
KW	Kuwait	PK	Pakistan	UY	Uruguay
KY	Cayman Islands	PL	Poland		
		PT	Portugal	VA	Vatican City (Holy See)
		PY	Paraguay	VC	Saint Vincent and the Grenadines
LA	Laos				
LB	Lebanon	QA	Qatar	VE	Venezuela
LC	Saint Lucia			VG	British Virgin Islands
LI	Liechtenstein	RO	Romania		
LK	Sri Lanka	RW	Rwanda	VN	Vietnam
LR	Liberia			VU	Vanuata
LS	Lesotho	SA	Saudi Arabia		
LU	Luxembourg	SB	Solomon Islands	WS	Samoa
LY	Libya	SC	Seychelles		
		SD	Sudan	YD	Democratic Yemen
MA	Morocco	SE	Sweden	YE	Yemen
MC	Monaco	SG	Singapore	YU	Yugoslavia
MG	Madagascar	SH	Saint Helena		
ML	Mali	SL	Sierra Leone	ZA	Republic of South Africa
MN	Mongolia	SM	San Marino	ZM	Zambia
MR	Mauritania	SN	Senegal	ZR	Zaire
MS	Montserrat	SO	Somalia	ZW	Zimbabwe

In 1988, Chemical Abstracts Service abstracted 80 795 chemical patent documents issued by 29 patent offices [14.1]. These are first cases (Chap. 15 and Section 19.2.1) covering the majority of all chemical inventions published.

Almost 54% of the chemical patent documents are issued by the Japanese Patent Office. The shares of the EPO and the USPTO are 10.3% and 6.9%, respectively.

The annual output of patent documents has been significantly enhanced by the introduction of deferred examination in the 1960s, resulting in publishing unexamined patent applications, and by the establishment of the patent systems of the PCT and the European Patent Convention, both publishing unexamined patent applications. These supranational patent systems did not replace the national patent systems but supplemented them.

15 Features Distinguishing Patent Literature from Scientific Literature

Patent literature has a number of special features that distinguish it from scientific literature and, in some respects, facilitate its use. Patent literature is a well-organized branch of technical literature. In each patent system it is issued by just one authority, the patent office, in constant cycles (usually weekly). There is no „gray" patent literature. In their patent gazettes and registers, the patent offices hold a comprehensive, up-to-date survey of the documents they have issued at the disposal of the users.

Patent documents never go out of print. Even very old patent documents are readily available, albeit as photocopies. The patent literature of many patent systems is classified according to a worldwide classification system, the *International Patent Classification*, IPC (Section 19.3.3). Currently (January, 1991) 75 patent offices apply the IPC which covers all sectors of technology. This makes it fairly easy to monitor, select, or purchase the patent documents pertaining to a specified field of interest (Chap. 17).

Format, layout, and the arrangement of the contents of patent documents (Chap. 16) have been harmonized to some extent in recent years. Here, again, the WIPO has done an extremely good job for the benefit of the readers of patent literature.

The patent documents issued in a patent system are uniformly priced, irrespective of their number of pages. The British patent specification no. 1 108 800, for example, comprises three volumes containing a total of 2055 pages and cost, when issued, only £ 1.

There is no copyright on patent documents [15.1], copying patent literature is not illegal.

In order to be fully informed on recent patent literature it is not necessary to read all published documents of all patent systems relevant to a certain sector of technology. The bibliographic data displayed on the title pages of patent documents (Section 16.2) lend themselves to grouping first-filed and follow-up documents into patent families (Section 19.2.1). Reading may then be confined to the first-issued document of each patent family, this reduces the reading effort to less than half.

Patent literature is abstracted in a timely, comprehensive, and reliable manner by commercial patent information services (Chap. 20). Reading the patent abstracts, instead of the entire patent documents, considerably reduces the workload of monitoring patent literature.

These features make patent literature comparatively easy to handle at the acquisition and current awareness stages, in spite of its inherently complex technical contents. Its complexity, on the other hand, strongly increases the effort required at the indexing and retrieval stages (Section 19.3).

16 Contents and Layout of Patent Documents

16.1 Introduction

Until the 1960s, the formats and layouts of patent documents mirrored the varieties of national patent systems. In a remarkable achievement the patent offices under the guidance of the WIPO have gradually adapted their patent documents to international standards. Many changes had to be made and the appearances of the documents have changed several times since the 1960s. Although changes still occur, harmonization has largely been completed. The *WIPO standards* not only pertain to the presentation of data and text but also to the sequence of the components (title page, claims, description of the invention, drawings and diagrams) of patent documents [16.1]. Their acceptance by national and supranational patent offices has greatly facilitated the handling, readability, and treatment of patent documents in information management.

16.2 The Title Page

The title pages of patent documents are added by the patent authorities to the texts filed by the applicants. They contain a wealth of data and are a source of valuable information relevant to patent information management. This is exemplified by the title pages of a European patent application (Fig. 7) and a European patent specification (Fig. 9).

The bibliographic data are each assigned a unique numerical code, the *INID Code* (INID = Internationally agreed Numbers for the Identification of Data, Figure 12). The INID Code numbers facilitate the identification of bibliographic data in documents written in languages the reader is not familiar with. Important bibliographic data on title pages are:

**INID CODES AND MINIMUM REQIRED FOR THE
IDENTIFICATION OF BIBLIOGRAPHIC DATA**

(10) **Document identification**

    *(11)   Number of the document

    *(12)   Plain language designation of the kind of document

    *(13)   Kind of document code according to WIPO Standard ST.16

   **(19)   WIPO Standard ST.3 Code, or other identification, of the office publishing the document

   Notes: (i)   ** Minimum data elements for patent documents only.

                  (ii) with the provisio that when data coded (11) and (13), or (19), (11) and (13), are used together and on a single line, category (10) can be used, if so desired.

(20) **Domestic filing data**

    *(21)   Number(s) assigned to the application(s), e.g. "Numéro d'enregistrement national", "Aktenzeichen"

    *(22)   Date(s) of filing application(s)

    *(23)   Other date(s), including date of filing complete specification following provisional specification and date of exhibition

    (24)   Date from which industrial property rights may have effect

    (25)   Language in which the published application was originally filed

    (26)   Language in which the application is published

(30) **Priority data**

    *(31)   Number(s) assigned to priority application(s)

    *(32)   Date(s) of filing of priority application(s)

    *(33)   WIPO Standard ST.3 Code identifying the national patent office allotting the priority application number or the organization allotting the regional priority application number; for international applications filed under the PCT, the Code "WO" is to be used

    (34)   For priority filings under regional or international agreements, the WIPO Standard ST.3 Code identifying at least one country party to the Paris Union for which the regional or international application was made

   Notes: (i)   with the provisio that when data coded (31), (32) and (33) are used together and on a single line, category (30) can be used, if so desired. If an ST.3 Code identifying a country for which a regional or international application was made is published, it should be identified as such using INID Code (34) and should be on a line separate from that of elements coded (31), (32) and (33) or (30)

                  (ii) The presentation of priority application numbers should be as recommended in WIPO Standards ST.10/C and ST.34.

(40) **Date(s) of making available to the public**

   **(41)   Date of making available to the public by viewing, or copying on request, an <u>unexamined</u> document, on which no grant has taken place on or before the said date

   **(42)   Date of making available to the public by viewing, or copying on request, an <u>examined</u> document, on which no grant has taken place on or before the said date

Figure 12. INID codes as defined by the World Intellectual Property Organization (Standard ST. 9) Reprinted by permission of the publishers

Figure 12. (Continued)

**(43)   Date of publication by printing or similar process of an <u>unexamined</u> document, on which no grant has taken place on or before the said date

**(44)   Date of publication by printing or similar process of an <u>examined</u> document, on which no grant or only a provisional grant has taken place on or before the said date

**(45)   Date of publication by printing or similar process of a document on which grant has taken place on or before the said date

  (46)   Date of publication by printing or similar process of the claim(s) only of a document

**(47)   Date of making available to the public by viewing, or copying on request, a document on which grant has taken place on or before the said date

Note: **Minimum data elements for patent documents only, the minimum data requirement being met by indicating the date of making available to the public the document concerned

(50) Technical information

  *(51)   International Patent Classification

   (52)   Domestic or national classification

   (53)   Universal Decimal Classification

  *(54)   Title of the invention

   (55)   Keywords

   (56)   List of prior art documents, if separate from descriptive text

            Note: Attention is drawn to WIPO Standard ST.14 in connection with the citation of references on the front page of patent documents and in search reports attached to patent documents.

   (57)   Abstract or claim

   (58)   Field of search

(60) References to other legally or procedurally related domestic patent documents including unpublished applications therefor

  *(61)   Number and, if possible, filing date of the earlier application, or number of the earlier publication, or number of earlier granted patent, inventors' certificate, utility model or the like to which the present document is an addition

  *(62)   Number and, if possible, filing date of the earlier application from which the present document has been divided out

  *(63)   Number and filing date of the earlier application of which the present document is a continuation

  *(64)   Number of the earlier publication which is "reissued"

   (65)   Number of a previously published patent document concerning the same application

   (66)   Number and filing date of the earlier application of which the present document is a substitute, i.e., a later application filed after the abandonment of an earlier application for the same invention

Notes: (i)   Priority data should be coded in category (30).

           (ii) Code (65) is intended primarily for use by countries in which the national laws require that re-publication occurs at various procedural stages under different publication numbers and these numbers differ from the basic application numbers.

Figure 12. (Continued)

(70) **Identification of parties concerned with the document**

  ** (71)   Name(s) of applicant(s)

    (72)   Name(s) of inventor(s) if known to be such

  ** (73)   Name(s) of grantee(s)

    (74)   Name(s) of attorney(s) or agent(s)

  ** (75)   Name(s) of inventor(s) who is (are) also applicant(s)

  ** (76)   Name(s) of inventor(s) who is (are) also applicant(s) and grantee(s)

  Notes: (i)   **For documents on which grant has taken place on or before the date of making available to the public, and gazette entries relating thereto, the minimum data requirement is met by indicating the grantee, and for other documents by indicating the applicant.

            (ii) (75) and (76) are intended primarily for use by countries in which the national laws required that the inventor and applicant are normally the same. In other cases (71) or (72) or (71), (72) and (73) should generally be used.

(80) **Identification of data related to International Conventions other than the Paris Convention**

    (81)   Designated states according to the PCT

    (83)   Information concerning the deposit of microorganisms, e.g. under the Budapest Treaty

    (84)   Designated contracting states under regional patent conventions

    (85)   Date of fulfillment of the requirements of Articles 22 and/or 39 of the PCT for introducing the national procedure according to the PCT

    (86)   Filing data of the regional or PCT application, i.e., application filing date, application number, and, optionally, the language in which the published application was originally filed

    (87)   Publication data of the regional or PCT application, i.e., publication date, publication number, and, optionally, the language in which the application is published

    (88)   Date of deferred publication of the search report

    (89)   Document number, date of filing, and country of origin of the original document according to the CMEA Agreement on Mutual Recognition of Inventors' Certificates and other Documents of Protection for Inventions

  Notes: (i)   The codes (86) and (87) are intended to be used:

               - on national documents when identifying one ore more of the relevant filing data or publication data of a <u>regional</u> or <u>PCT</u> application, or

               - on <u>regional</u> documents when identifying one ore more of the relevant filing data or publication data of <u>another regional</u> or <u>PCT</u> application.

            (ii) all data in code (86) should be presented together and preferably on a single line.

            (iii) all data in code (87) should also be presented together and preferably on a single line.

1) The double-digit *country code* (INID Code number 11) and the document number (*patent number*) are used for unequivocal identification of the document. The country codes are listed in Table 11. The patent numbers are formed according to rules that vary widely between patent systems. Some patent offices have adopted entirely new numbering systems. The German Patent Office has recently changed its numbering system.

   There are two main ways of numbering patent documents. *Chronological numbering at acceptance* of patent applications: When the numbers allocated are carried through all stages of publication (application, specification, revised specification), the sequence of numbers is not complete because some patent applications are withdrawn or rejected. When examination is deferred, the patents are not issued in an increasing number sequence because some applicants may wait longer than others before they initiate examination; some may even refrain from examination at all (e.g., in Germany and Japan). *Chronological numbering at publication* either of patent applications or of patent specifications is also practiced, e.g., by the USPTO. Some patent offices use two-tier systems which are mixtures of both approaches, e.g., the French Patent Office (INPI).

2) The *type of document code* (INID Code number 13; A1 in Fig. 7, INID Code not allocated) indicates the legal status. „A" designates a patent application. A patent specification is denoted by „B"; the specification issued at grant has the code B1 (Fig. 9), a revised patent specification has B2 (Fig. 11). The code A1 means that the application has been published together with the corresponding search report. An application published without its search report is denoted by the code A2. The delayed search report is issued together with the title page under the code A3. Other patent systems use other type of document codes reflecting their different application and granting procedures.

3) INID Code number 21 designates the number under which the application has been accepted (*application number*). In the European patent system, the nine-digit application number is a combination of four data elements. Its first two digits are formed by the last two digits of the year of application (89 = 1989). The third and fourth digits are a code for the office with which the application was filed (30 = United Kingdom). The next four digits form the chronological number of the application allocated at the accepting office. The last digit, preceded by a full stop, is a control digit. In other patent systems, application numbers follow different rules. The importance of application numbers for creating patent families is explained in Section 19.2.1.

4) The *filing date* of the application is marked by INID Code number 22. In most patent systems the term of a patent starts on this date.

5) INID Code number 30 indicates the *priority data*: date(s) of first filing and number(s) of the application(s) for which priority is claimed (priority application number(s)). The priority data are relevant to patent information management with respect to patent families (Section 19.2.1) and to prior art searches (Section 19.3).

6) The *publication date* and the *issue number of the patent gazette* in which the publication is announced are indicated by INID Code number 43. The publica-

tion date is important when this application is considered as prior art against other patents.

7) The pertinent *classes of the International Patent Classification* (IPC, Section 19.3.3) or, if a national classification is used instead of or in addition to the IPC (as in the United States), the classes of that classification, are indicated by INID Code number 51 or 52. INID Code numbers 54 and 56 respectively are allocated to the *title of the application* and *prior art literature* cited by the patent office. When opposition is considered against the patent it is always wise to study the cited prior art carefully. INID Code number 57 is associated with the *abstract* of the disclosure prepared by the applicant.

8) The *name(s) of the applicant(s) and the inventor(s)* (INID Code numbers 71 and 72) are useful for monitoring competitors' patent activities (Section 19.2).

9) INID Code numbers 84 to 86 are related to the *designated states* and other information specific to the European or PCT patent systems. The *deferred publication of the search report* is given under INID Code number 88. Substantive examination must be applied for within 6 months from this date .

10) At the bottom of the title page of a European patent specification (Fig. 9) the public is reminded of the nine months' opposition period.

16.3   The Description of the Invention

The requirements of the patent laws for the patentability of inventions are dealt with in Section 12.2 and Chapter 13. The applicant is well advised to disclose his invention clearly, intelligibly (to the specialist), in full accordance with the facts and comprehensively, otherwise he might run the risk that his application is rejected or his patent revoked. If he intends to file follow-up applications in other patent systems he must bear in mind that the clarity and comprehensiveness of his description will be judged from different viewpoints in different patent systems.

The description of the invention usually starts with the *prior art* known to the inventor and the advantages of his invention in relation to this prior art. For readers not fully familiar with a certain sector of technology, the introductory part of a description offers a valuable piece of technological information.

The description then gives a detailed explanation of the invention. Often, a number of workable examples or „recipes" and their results are presented. If necessary, the description is completed by schemes, drawings, or chemical structure diagrams.

16.4   The Patent Claims

A patent claim is a concise description of the essential features of the invention. It is a legally binding definition of the product or process for which protection is sought. A patent document may contain more than one claim, possibly referring to different aspects of the invention (e.g., a product and a process of its manufacture), or the different claims may be arranged by increasing specificity.

In many patent sytems patents protecting chemical substances can be obtained. These composition-of-matter claims provide the broadest protection possible because they also cover any preparation or application of that particular compound or compounds.

Claimed chemical compounds are usually expressed in terms of structural formulae. When a claim covers a whole class of chemical compounds with a common structural feature (substructure) this class is usually defined by a *generic formula*, sometimes (wrongly) called *Markush formula* [16.2].

Generic formulae contain one or more „fixed" substructures with variable substitution or linkage, the variations being generated by lists of alternative substituents or substructures with fixed or variable chain lengths or ring sizes, on fixed or variable positions or in variable numbers etc. Figure 13 shows a generic formula. Generic formulae may comprise thousands or even millions of chemical compounds and pose specific difficulties with respect to both their interpretation and indexing and to information retrieval. These problems are discussed extensively in the literature (Section 19.3.4).

When a patent infringement suit comes to court, the scope of protection is determined on the basis of the patent claim(s), interpreted in the light of the description of the invention. Obviously, careful wording of patent claims is essential to obtain adequate protection. The formulation of patent claims requires experience and skill. Complex language and involved logic are frequently encountered because the value of a patent, its ability to stand up to litigation and the economic reward for disclosing the invention to competitors depend heavily on the wording of the patent claims.

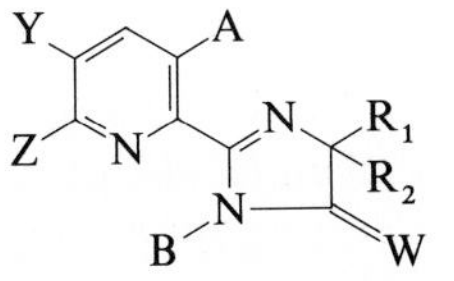

$R_1 =$	1–4C alkyl
$R_2 =$	3–6C cycloalkyl, optionally substituted by Me
$A =$	$COOR_3$ or $CONHR_6$
$R_3 =$	di(lower alkyl)methylamino or a cation
$R_6 =$	3C alkenyl, 3–16C alkynyl, optionally substituted by OH or Cl
$B =$	COR_4 or So_2R_5
$R_4 =$	1–11C alkyl or phenyl, optionally substituted by Cl, NO_2 or OMe
$R_5 =$	1–4C alkyl, phenyl or tolyl
$W =$	O or S

Z and Y = 1–6C alkyl or phenyl, optionally substitued by 1–4C alkyl or 1–4C alkoxy

Figure 13. A generic structural formula

17   Availability, Monitoring, Forms, and Storage of Patent Documents

17.1   Authorities and Services Supplying Patent Documents

Patent documents are available from the distribution services of the patent offices and from commercial services. The patent documents of patent systems with two-tier numbering must be ordered by publication number.

17.2   Monitoring and Selecting Patent Documents

As patent documents are published in complete, well-ordered series in a fixed sequence, and are clearly identified by their numbers, a patent office's output is easy to monitor. Patent documents can be purchased by IPC classes; subscriptions to individual classes can be made.

Industrial enterprises maintain collections of patent documents. The documents may serve as a basis of input of bibliographic or technical data, as reference material for the patent specialists, they may be circulated among research and development staff for current awareness, they may be used for checking the relevance of computer searches and be delivered to users as output of patent information retrieval.

It is neither reasonable nor feasible to collect the whole range of published patent documents which might be of interest to a company. A choice must be made as to which documents (published patent applications, patent specifications, utility model descriptions, etc.), of which patent systems, and which sectors of technology are to be collected for filing in the corporate patent document library. This decision should always be made in accordance with the patent and the research and development departments of the company. The countries in which the company's and its competitors' major centers of production and sales are located have to be taken into account.

Information specialists need the first-published documents of the respective patent families (first cases, Section 19.2.1), generally the published patent applications, as raw material for their input. The patent specialists must monitor the patent specifications which display the patent claims as allowed by the examiners to decide on opposition procedures, litigations, or taking licenses.

The technology sectors that are relevant to the company's business can be expressed by IPC classes. As the IPC is not applied consistently in the patent offices (Section 19.3.3), it is also advisable to monitor IPC classes that appear to be outside the company's focus of interest.

The patent gazettes or other suitable organs (Section 18.3) must be monitored continually according to the above criteria for selecting patent documents to be ordered.

17.3   Patent Documents on Microforms and CD-ROMs

Patent documents have been available for many years on paper (printed, photocopies, or photostat copies). The maintenance of large collections of paper documents (Section 17.4) poses great problems. Patent offices and commercial services therefore started to offer patent literature in microforms. Some patent offices have even discontinued issuing paper documents. Microforms drastically reduce storage space, but render the access to the documents cumbersome.

Reeled microfilm is the cheapest form, but is only useful for storing older patent documents that are not frequently used. Readers and copiers for reeled microfilm equipped with automatic retrieval devices are expensive and can only be used for films bearing optical markers (blips) which are not available for the patent literature in all countries. Moreover, the scale of reduction and the arrangement of the frames on the films are not standardized which means that different equipment may be required for the microfilm documents from different countries.

Microfiches are even less suitable for machine handling and do not have any advantages over reeled microfilm. Unfortunately, some patent offices have discontinued producing reeled microfilm in favor of microfiche. A viable alternative might be to purchase reeled microfilm from commercial services that cover a wide range of patent systems and years of issue. This can make the handling of microforms more uniform.

Obviously, the use of patent literature is heaviest for the most recent years of issue. Thus, for current awareness and input purposes patent documents on paper cannot be substituted by microforms (optical storage media may however be a suitable alternative, see below). Another kind of microform has gained some popularity for purposes requiring only occasional access to copies: patent documents on *aperture cards* which are issued, for example, by the European, German, French and Belgian patent offices. Aperture cards are punch cards that have a microimage insert displaying eight pages of a patent document each. They did not prove to be useful for their original purpose (i.e., sorting by punchcard sorting machines) because the cards were soon torn up when heavily used. However, they are more convenient to handle than reeled microfilm or microfiche when used as card file packages; the individual aperture cards are copied automatically from the packages by special equipment [17.1]. However, this equipment is relatively expensive, and aperture cards might be economical only at high throughput. Another problem has been the varying quality of the microimages of the aperture cards, particularly those produced by the EPO. A new era of patent document handling dawned when the EPO started to issue its patent documents on CD-ROMs in 1989 [17.2]. Three types of these exceptionally low-priced CD-ROMs are distributed:

1) *ESPACE* comprises the complete sets of A1 and A2 patent applications, including drawings and diagrams, from 1989 onwards. About 60 CD-ROMs are issued annually. Thirteen different data elements can be retrieved by the software stored on the optical discs. The documents are stored in facsimile format and produce high-quality copies on laser printers.
2) *WORLD* started in 1990 and contains the PCT applications issued by the WIPO with 500 applications per CD-ROM and the same retrieval facilities as ESPACE.
3) *FIRST* contains the images of the title pages only of A1 and A2 European applications (5 and 6 CD-ROMs for 1988 and 1989, respectively) with the data retrievable as in ESPACE. From 1990 onwards, the European title pages are combined with the PCT title pages as *FIRST EPA/PCT*.

Provided that workstations equipped with CD-ROM drives, high-resolution monitors, and fast laser printers are at hand, a patent document collection may be run on CD-ROM. Under certain conditions (frequent use, high number of copies to be provided, etc.), this method can be economical and fully substitute the purchase of printed patent documents. There are, however, some inherent problems. First, for the time being (spring 1991) the EPO is the only patent office issuing patent documents on CD-ROMs. Other patent offices have announced that they will follow suit, but it will take some time until all major patent offices have done so. It is to be hoped that they all comply with the same standard! For the next 5–10 years, information managers will probably therefore have to handle patent documents both by traditional methods and by using the new storage media.

A second problem is the number of CD-ROMs accumulating over the years. The EPO will issue 60 CD-ROMs for ESPACE alone. Within a relatively short period this output will build up a CD-ROM collection that can no longer be handled manually. Additional CD-ROMs from other patent offices will quickly enhance the predicament. Disc storage devices (*juke boxes*) for many CD-ROMs are on the market but it is doubtful that they will solve the storage and retrieval problems at reasonable cost. An alternative is the storage of the contents of the CD-ROMs on WORMs or similar gigabyte optical storage media. That would open up novel opportunities of making free and ample use of the text and data of patent documents (e.g., by transmitting them to the scientists' desk-top computer monitors). This, again, requires high investments in computer and telecommunication hardware and software.

A third problem is that the CD-ROMs always contain the entire series of patent documents issued. This material is more than most customers are interested in and is far too much for small- and medium-sized enterprises or patent attorneys. Of course, a patent office cannot be expected to produce selections of patent documents on CD-ROMs tailored to the individual needs of customers. In view of the high cost of CD-ROM mastering, this would jeopardize the economy of publishing patent documents on CD-ROMs. Thus, the use of CD-ROMs in patent information management may remain an option which can only be fully and economically exploited by big companies or organizations. Nevertheless, the EPO has pioneered a development that should be closely watched by the information profession.

17.4   Storage of Patent Documents

As IPC classes are printed on the title pages of patent documents, paper documents can be arranged as classified files for manual searches. This is still the preferred approach of the patent offices for maintaining their search files. The disadvantage is that these files require a great deal of space, particularly when the documents are not only arranged by main classes but also by the additional classes printed on the title pages. The effort required to keep these comprehensive paper files up to date is tremendous.

In an industrial environment large collections of patent literature should be used in conjunction with information retrieval systems. The resulting hit lists then guide the searcher to the positon of the documents in the archives, which can simply be kept in numerical order. Current developments in patent information management have lead to computer searches in databases (Section 19.3) and storage of patent literature on paperless media (Section 17.3), so that manual searches in patent document files have gradually become obsolete.

18   Patent-Related Literature

18.1   Patent Gazettes

In addition to regularly issued patent documents, each patent office publishes a patent gazette listing the patent applications laid open to public inspection, the patents granted and any other patent-related transactions that have occurred in that particular week. These transactions become legally valid with their notification in the patent gazette; therefore, careful cognizance of the patent gazette entries is mandatory to patent information management. Figure 14 shows the table of contents of the European Patent gazette (*European Patent Bulletin*), two entries of the European patent gazette are shown in Figures 8 and 10.

Here, as in some other patent gazettes, the entries are confined to the essential bibliographic data, so that the subjects of the inventions can only be inferred from the titles and the IPC classes. This is not a reliable basis for selecting relevant patent documents (Section 17.2). Other publication media (Section 18.4) may be used instead.

The gazettes of other patent systems are more informative, they display abstracts (e.g., France and PCT) or the main claims (e.g., United States), together with drawings or structural diagrams. Figure 15 shows a page taken from the patent gazette of the USPTO (*Official Gazette*).

Sometimes, patent documents are published with a delay. This is legally irrelevant. The opposition period for example begins with the announcement of the grant of the patent in the patent gazette, even if the patent specification is issued much later or not at all.

18.2   Official Journals and Communications

It is the duty of a patent information manager to take note of official announcements and other important events related to intellectual property protection. In some patent systems these are communicated in the patent gazettes (e.g., in Austria, the United Kingdom, the United States, Switzerland), in others they appear in separate publications. Examples are the *Official Journal of the European Patent Office* (texts in German, English, and French) and the *Blatt für Patent-, Muster- und Zeichenwesen* edited by the German Patent Office. In addition to patent office announcements, these publications contain statistical data, important decisions of the patent office or of courts, texts of new laws, and other relevant pieces of infor-

(31/1991) 31.07.1991 Europäisches Patentblatt
European Patent Bulletin
Bulletin européen des brevets **A**

A Inhalt A Contents A Sommaire

Figure 14. Table of contents of the European Patent Bulletin

Europäisches Patentblatt
European Patent Bulletin
Bulletin européen des brevets

A   **(31/1991) 31.07.1991**

II.	Erteilte Patente	269	II.	Granted patents	269	II.	Brevets délivrés	269
II.1	Geordnet nach der Internationalen Patentklassifikation	269	II.1	Arranged in accordance with the International patent classification	269	II.1	Classés selon la classification internationale des brevets	269
II.2 (1)	Geordnet nach Veröffentlichungsnummern	327	II.2 (1)	Arranged by publication number	327	II.2 (1)	Classés selon les numéros de publication	327
II.2 (2)	Geodnet nach Anmeldenummern	331	II.2 (2)	Arranged by application number	331	II.2 (2)	Classés selon les numéros de dépôt	331
II.3	Geordnet nach Namen der Patentinhaber	335	II.3	Arranged by name of proprietor	335	II.3	Classés selon les noms des titulaires	335
II.4	Frei	—	II.4	Reserved	—	II.4	Réservé	—
II.5	Tag des Erlöschens des europäischen Patents in einem Vertragsstaat	343	II.5	Date of lapse of the European patent in a Contracting State	343	II.5	Date de déchéance du brevet européen dans un Etat contractant	343
II.6 (1)	Einspruch	343	II.6 (1)	Opposition	343	II.6 (1)	Opposition	343
II.6 (2)	Kein Einspruch eingelegt	346	II.6 (2)	No opposition filed	346	II.6 (2)	Il n'a pas été fait opposition	346
II.7 (1)	Einspruch, der als nicht eingelegt gilt	355	II.7 (1)	Opposition deemed not to have been filed	355	II.7 (1)	Opposition réputée non formée	355
II.7 (2)	Unzulässiger Einspruch	355	II.7 (2)	Inadmissible opposition	355	II.7 (2)	Opposition non recevable	355
II.7 (3)	Widerruf des europäischen Patents	355	II.7 (3)	Revocation of the European patent	355	II.7 (3)	Révocation du brevet européen	355
II.7 (4)	Aufrechterhaltung des europäischen Patents in geändertem Umfang	355	II.7 (4)	Maintenance of the European patent as amended	355	II.7 (4)	Maintien du brevet européen dans sa forme modifiée	355
II.7 (5)	Zurückweisung des Einspruchs (der Einsprüche)	356	II.7 (5)	Rejection of the opposition(s)	356	II.7 (5)	Rejet de l'opposition (des oppositions)	356
II.7 (6)	Einstellung des Einspruchsverfahrens	356	II.7 (6)	Termination of opposition procedure	356	II.7 (6)	Clôture de la procédure d'opposition	356
II.8 (1)	Tag des Eingangs des Antrags auf Wiedereinsetzung in den vorigen Stand	356	II.8 (1)	Date of receipt of request for re-establishment of rights	356	II.8 (1)	Date de réception de la requête en rétablissement dans un droit	356
II.8 (2)	Tag und Art der Entscheidung über den Antrag auf Wiedereinsetzung in den vorigen Stand	356	II.8 (2)	Date und purport of decision on request for re-establishment of rights	356	II.8 (2)	Date et type de la décision relative à la requête en rétablissement dans un droit	356
II.9 (1)	Tag der Aussetzung im Fall der Regel 13	356	II.9 (1)	Date of suspension in the case referred to in Rule 13	356	II.9 (1)	Date de la suspension dans le cas de la règle 13	356
II.9 (2)	Tag der Fortsetzung im Fall der Regel 13	356	II.9 (2)	Date of resumption in the case referred to in Rule 13	356	II.9 (2)	Date de la poursuite dans le cas de la règle 13	356
II.9 (3)	Tag der Unterbrechung im Fall der Regel 90	356	II.9 (3)	Date of interruption in the case referred to in Rule 90	356	II.9 (3)	Date de l'interruption dans le cas de la règle 90	356
II.9 (4)	Tag der Wiederaufnahme im Fall der Regel 90	356	II.9 (4)	Date of resumption in the case referred to in Rule 90	356	II.9 (4)	Date de la reprise dans le cas de la règle 90	356
II.10	Frei	—	II.10	Reserved	—	II.10	Réservé	—
II.11	Frei	—	II.11	Reserved	—	II.11	Réservé	—
II.12	Änderungen	356	II.12	Changes	356	II.12	Modifications	356
II.13	Berichtigungen	359	II.13	Rectifications	359	II.13	Rectifications	359

Figure 14. (continued)

mation. The monthly communications bulletin of the WIPO, *Industrial Property*, is also worth reading.

18.3   Other Patent-Related Literature

Defensive publications constitute a special kind of patent-related literature that intentionally do not lead to monopolies. They are published to prevent others from

produce a methyl 4-O-protected-2,6-dideoxy-2-fluoro-α-L-talopyranoside of the formula

(V)

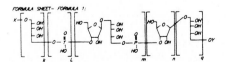

wherein B is as defined above, and
(d) removing the hydroxyl-protecting group (B) from the compound of the formula (V) to produce the compound of the formula (Ia).

5,034,518
2-FLUORO-9-(2-DEOXY-2-FLUORO-β-D-ARABINOFURANOSYL) ADENINE NUCLEOSIDES
John A. Montgomery, and John A. Secrist, III, both of Birmingham, Ala., assignors to Southern Research Institute, Birmingham, Ala.
Filed May 23, 1989, Ser. No. 355,358
Int. Cl.^{5} C07H *19/19, 19/16*
U.S. Cl. 536—26   2 Claims
1. A nucleoside having Formula I

I

wherein R is hydrogen or acyl.

5,034,519
OLIGOSHACCHARIDES, IMMUNOGENS AND VACCINES, AND METHODS FOR PREPARING SUCH OLIGOSACCHAIDES, IMMUNOGENS AND VACCINES
Eduard C. Beuvery, Vianen; Adolf Evenberg, Utrecht; Jan T. Poolman, Broek in Waterland; Jacobus H. van Boom, Voorschoten; Peter Hoogerhout, Gouda, and Constant A. A. van Boeckel, Oss, all of Netherlands, assignors to De Staat Der Nederlanden, Netherlands
Filed Dec. 29, 1987, Ser. No. 139,349
Claims priority, application Netherlands, Dec. 31, 1986, 86.03325
Int. Cl.^{5} C07H *13/00, 1/00;* C08B *37/00*
U.S. Cl. 536—117   9 Claims

FORMULA SHEET- FORMULA 1:

1. A synthetic Oligosaccharide which comprises D-ribose, D-ribitol and phosphate in any order repeated m times, wherein m is an integer from 2 to 20.

5,034,520
PROCESS FOR RECOVERING HEPARINIC OLIGOSACCHARIDES WITH AN AFFINITY FOR CELL GROWTH FACTORS
Jean-Claude Lormeau, Maromme; Maurice Petitou, and Jean Choay, both of Paris, all of France, assignors to Sanofi, Paris, France
Filed Apr. 16, 1987, Ser. No. 39,471
Claims priority, application France, Apr. 17, 1986, 86 05546
Int. Cl.^{5} C07H *1/00*
U.S. Cl. 536—127   12 Claims
1. A process for obtaining high anionic strength heparinic oligosaccharides from a mixture of heparinic glycosaminoglycans, which mixture of heparinic glycosaminoglycans is substantially free of non-heparin components and is produced by the depolymerization of heparin, comprising the steps of:
   contacting the mixture of glycosaminoglycans with a cationic or anionic cell growth factor to bind said oligosaccharide to the growth factor; and then
   eluting said heparinic oligosaccharides from the growth factor.

5,034,521
PROCESS FOR PREPARING 3-SUBSTITUTED METHYL-3-CEPHEM-4-CARBOXYLIC ACID
Kentaro Fukuzaki, and Wataru Takahashi, both of Nobeoka, Japan, assignors to Asahi Kasei Kogyo Kabushiki Kaisha, Osaka, Japan
Filed May 22, 1989, Ser. No. 356,034
Claims priority, application Japan, May 24, 1988, 63-124879
Int. Cl.^{5} C07D *501/04*
U.S. Cl. 540—230   7 Claims
1. A process for preparing a 3-substituted methyl-3-cephem-4-carboxylic acid represented by formula (I) or a pharmaceutically acceptable salt thereof

(I)

wherein R^1 represents a hydrogen atom or a lower alkoxy group and R^2 represents a lower alkyl group unsubstituted or substituted with a halogen atom; a nitro group; an alkoxy group having 1 to 6 carbon atoms; an alkylthio group having 1 to 6 carbon atoms; an alkylamino group having 1 to 6 carbon atoms; a dialkylamino group having 2 to 12 carbon atoms; an acylamino group having 1 to 6 carbon atoms; or an acyl group having 1 to 6 carbon atoms, or an aryl group selected from the group consisting of a phenyl group, a tolyl group, a xylyl group, a benzyl group and a phenethyl group, said aryl group being unsubstituted or substituted wth a halogen atom; a nitro group; an alkoxy group having 1 to 6 carbon atoms; an alkylthio group having 1 to 6 carbon atoms; an alkylamino group having 1 to 6 carbon atoms; a dialkylamino group having 2 to 12 carbon atoms; an acylamino group having 1 to 6 carbon atoms; or an acyl group having 1 to 6 carbon atoms, which comprises reacting a cephalosporanic acid represented by formula (II) or a pharmaceutically acceptable salt thereof

(II)

wherein R^1 has the same meaning as defined above, with a compound represented by formula (III)

Figure 15. A page taken from the Official Gazette

obtaining a patent on an idea which an inventor or his employer feels is not worth patenting or should not be patented for specific reasons. The idea may be divulged in any publication (e.g., a technical journal), but the contents and style of a defensive publication are not always acceptable to the editors of journals.

Defensive publications lie in the border area between patent publications and scientific publications. Before deferred examination was introduced in the 1960s (first in the Netherlands), patent applications filed with the Belgian Patent Office, abandoned after they had been laid open to public inspection at a very early date, were used as defensive publications. Nowadays, a considerable proportion of defensive publications are found among the published unexamined patent applications.

In the United States, applications with a definitive defensive purpose (Defensive Publications, Statutory Invention Registrations) can be filed with the USPTO and are published in the *Official Gazette* (Fig. 15). Two journals, *Research Disclosure* [18.1] and *International Technology Disclosures* [18.2], have specialized in divulging defensive publications.

Apart from official communication bulletins, prominent journals related to intellectual property protection are issued by national professional organizations or commercial publishers. Examples are: *Gewerblicher Rechtsschutz und Urheberrecht* (GRUR), appearing in a national and an international edition [18.3], *Mitteilungen der deutschen Patentanwälte* [18.4] in Germany; *Journal of the Patent and Trademark Office Society* [18.5] in the United States; *World Patent Information* in the United Kingdom [18.6]. Patent information managers should not miss reading these.

An interesting series of publications, the *Auszüge*, are issued by a German publishing house [18.7]. The weekly outputs of the German and European patent offices (patent applications, patent specifications, utility model descriptions) are reported by their bibliographic data, main claims, drawings and diagrams in IPC class order. The brochures are issued weekly, shortly after the patent documents are available. The *Auszüge* are extremely useful for early current awareness. They contain much more information than the European and German patent gazettes and may be used for reliable monitoring of the European and German patent literature.

19 Principles, Methods, and Problems of Patent Information Management

19.1 Introduction

The principles governing patent information management, the methods employed, and the problems encountered in processing patent information are due to the two-tier function of patent literature (Chap. 13) and the diverse requirements of its user communities.

Patent specialists make particularly strong demands on the completeness of patent searches: Failing to notice relevant prior art before filing a patent application may result in the rejection of the application or in the revocation of a granted patent after an invention has already been disclosed to the public. The effort put into research and development leading to the patent application would then be fruitless, and competitors might needlessly have gained insight into the research and development policy of the applicant.

Being unaware of an existing patent of a competitor might lead to the infringement of this patent. The money invested in the preparations for producing and marketing a product might be lost and substantial damages may have to be paid to the patent owner. The legal and financial consequences of patent infringement can be disastrous [19.1].

Complete search results are difficult to achieve. The searcher has to be experienced and the search strategy must be chosen with utmost care. The reliability of a search depends not only on the searcher's skills; the outcome of a search is also influenced by the quality of the search file (e.g., with respect to coverage of the patent literature in the technological area in question), the power of the indexing and retrieval language, the exhaustivity, correctness and consistency of indexing, the time-lag between publication and searchability of the information, and other criteria. When searches are done in external data bases — commercial, public, or private — (Chap. 20) as is now commonplace, the problem is aggravated because most of these factors are beyond the searcher's control.

In principle, every error entering the search file in the long and complex process of producing a data base and making it available can endanger the reliability of a search. When this is understood and related to the vast amounts of data processed for and stored in existing patent data bases it becomes obvious that no data base can be flawless and complete. Numerous investigations have corroburated this statement [19.2]–[19.8].

Experienced searchers know that any two data bases covering the same area of technology will almost never render identical search results for the same query. Instead, the results will overlap: some citations are retrieved in both data bases, some only in one, some only in the other. To make patent searches as watertight as possible, it has therefore been common practice to search in more than one data base and to combine the results (*cross-file* or *cross-host searches*). In theory, the overall result will improve with every additional data base taken into account, but practical reasons set a limit to doing so: only few patent data bases of comparable coverage are available; the average searcher will hardly ever fully master more than three or four retrieval and command languages; because of the overlap, the proportion of irrelevant citations to be sifted out will increase steeply with every additional data base searched in. Therefore, multi-data base searches tend to be very time-consuming or even unworkable.

High numbers of irrelevant citations are typical for searches geared to complete recall because recall is inversely related to search precision. The specific nature of this relationship depends on the indexing and retrieval language investigated and has been a matter of discussion for years [19.9]–[19.11]. The increase in the number of irrelevant citations with increasing recall is worst in full-text data bases and least in data bases furnished with powerful indexing and retrieval languages (e.g., topological structure data bases, see Section 19.3.4).

Not every search in the patent literature is set on completeness. A research chemist is often content with an incomplete search result, but this is then expected to contain only citations relevant to the problem that forms the background of his query. The client will strongly object to being swamped with masses of irrelevant patent literature. In this case the search strategy may be narrowly focussed — as far as the respective retrieval language permits — to specific search terms. Because of the inverse relationship between recall and precision, this kind of search cannot and will never be complete.

As explained in Chapter 20, producers of chemical patent data bases have been aware of the recall/precision dilemma and have adjusted their products accordingly: the existing data bases may be grouped into those which are, in their indexing policy and their retrieval tools, best suited for high recall searches (*recall-type data bases*) and those which are more appropriate to precision searches (*precision-type data bases*). Thus, the searcher can make a choice according to his information needs.

19.2   Registration and Use of Bibliographic Patent Data

Bibliographic patent data as they are displayed on the title pages of patent documents (Section 16.2), in the patent gazettes (Section 18.1), and in the registers kept in the patent offices (Section 21.2) contain a great deal of information that must be processed and made retrievable.

The users' and the information specialists' interest in bibliographic patent data centers on four aspects: (1) identifying existing relationships between patents or patent applications, (2) determining the legal status of patents or patent applications,

(3) monitoring competitors' patent activities, and (4) discovering clusters of prior art by means of citations.

Moreover, bibliographic patent data are extremely useful as search criteria for limiting searches for technical disclosure to time ranges (priority or publication dates) or to certain applicants or inventors.

In cross-file and cross-host searches the citations retrieved in different data bases must be linked by their bibliographic data. This requires standardization of bibliographic patent data among the respective data bases and hosts [19.12], [19.13].

User groups and international associations (Section 22.2.2) have pointed out the great importance of bibliographic patent data for information management to patent offices (as the producers of bibliographic data) and to the databank producers (as the processors of bibliographic data), urging them to check the published or stored data more carefully and to improve the reliability of data searches.

19.2.1   Patent Families [19.14], [1915]

As explained in Section 12.2, patent protection for inventions is usually sought for in more than one patent system. On average, three patent applications are filed for each invention. The patents and patent applications belonging to the same invention, consisting of the first filed application plus the corresponding follow-up applications as well as the patents granted on them, form a *patent family*. The members of a patent family are *equivalent* to each other.

The published patent applications and patent specifications of a patent family are not issued at the same time, the durations of the publication and granting procedures in the patent offices differ. In some national patent systems (e.g., in the USA) patent specifications might be published several years after application. The patent systems in which patent applications are to be published unexamined 18 months after their application or priority dates (Section 12.2) also differ considerably in punctuality.

The question therefore arises as to how equivalent patent documents can be found together. Here, the priority data printed on the title pages (INID Code number 30, Section 16.2) provide a convenient solution because every patent family is earmarked by the same priority date and number of first-filed application (*priority application number*). This only holds true when priority is claimed for the follow-up applications, which is not so for about 3–5% of the applications. These *non-Convention cases* require special treatment (see below) if comprehensive registration of patent families is aimed at.

In reality, correlation of equivalent applications and patents is not as easy as it looks at first sight. Quite often, follow-up applications have more than one priority, and the "mix" of priorities might vary within a patent family in the follow-up applications filed in different patent systems. Multi-priority cases belong to as many patent families as there are priority application numbers cited under INID Code number 30. In the United States, an application may be refiled one or more times, as a continuation or continuation-in-part application (Section 12.1). Therefore, in a U.S. patent specification, whole series of priority application numbers and priority dates might be cited [19.16]. Multiple priorities create equivalence networks rather than one-to-one relationships.

Priority application numbers can be ambiguous. In the United States, for example, application numbers (*serial numbers*) have six digits, running from 000001 to 999999, whereas patent numbers have seven. The serial numbers start again with 000001 when the six-digit range is exhausted, thus, they are unambiguous only in combination with the year of application. These and other problems have to be considered in processing bibliographic patent data.

Not long ago, the allocation of incoming patent documents to patent families was a formidable and demanding task: priority details had to be collated and transferred to card files. Now, bibliographic patent data are available in computer-readable form from patent offices, and public and commercial patent information services (Chap. 20) process them and provide data banks for patent family searches.

By carrying out searches in these data banks, a variety of user requirements can be met: A research chemist might have become aware of an invention patented in, say, Japan; he wants to read the description and asks for an equivalent document written in a language he is familiar with. A patent specialist might inquire whether there is a patent or application in a certain patent system that is equivalent to a patent he knows. If he is involved in license negotiations, he might want a comprehensive survey over the patent portfolio of a company in a defined time range. Alternatively, if he wants to examine whether the priority of a competitor's application is justly claimed, he will need to know the other members of the patent family. Several other situations in which patent family searches play a role can be itemized.

Searches for equivalent patent documents are also indispensible in the context of processing patent information: Priority can only be claimed for a follow-up application if it is based on exactly the same invention as described in the first-filed application (Sections 12.1 and 12.3), i.e., if the descriptions of the invention are identical in both documents. It follows that every published patent document belonging to the same patent family is identical with respect to its description of the invention, provided that priority is claimed (whereas, affected by the laws and granting procedures of the patent systems, the claims may differ). This has the advantage that the analysis and the indexing of the technical disclosure of patent documents can be restricted to just one member of each patent family. This reduces the indexing effort to about one third and is an enormously cost-effective measure. As a rule, the first-to-publish document of a patent family (the *first case*), which is not always the first-filed case, is taken for indexing. Subsequently published documents are simply allocated to their patent families by means of their bibliographic data.

Patent data banks must be searchable from different access points. Priority application numbers, priority dates, applicants' and inventors' names and patent numbers should function as access keys. Names pose specific problems in data processing [19.17]–[19.18]. Translation or transliteration can heavily distort applicants' names, so that special computer programs must be employed for their standardization. When names are used as search keys in patent family data banks, it has to be taken into account that an applicant's name might vary within a patent family because the equivalent applications might have been filed by different affiliate companies.

A patent family data bank established by computer correlation and processing of priority information can only locate patent documents with identical priority application numbers. It therefore fails to identify follow-up applications or patents for which no priority has been claimed. Non-Convention cases can only be detected and allocated to their patent families by careful intellectual analysis of the patent documents on the basis of their technical disclosure.

Because patent specialists use patent data as a basis for legal actions, patent data bank entries have to be carefully checked both intellectually and by elaborate computer programs [13.7].

19.2.2   Legal Status Information

Not only information on patent families is relevant to the patent specialists. They often also want to know about the *legal status* of a patent or patent application, e.g., at what date the application was published, whether a patent has already been granted , and if so, when, whether a search has already been requested, and if so, what the result was.

Such questions can be answered by a case-by-case check of the patent registers (Section 21.2) or by using a file which is regularly fed with the relevant entries selected from the patent gazettes of the major patent systems. This has been a tedious and costly task and has become even more so since the supranational patent systems (Section 12.3) with their complex procedures and their multiple entanglements with national patent systems came into being. Fortunately, commercial and public information services have recently established *patent status data banks* that can be accessed online (Chap. 20).

A task that the patent specialists use to delegate to the information managers is the periodical monitoring of certain competitors' patent applications which might threaten the company's activities if they lead to patents. A company might have to watch many of these, and this cannot be efficiently done by case-by-case inspection of the patent gazettes or patent registers. Here, the commercial or public services maintaining patent status data banks can help to create a corporate patent monitoring system that automatically issues messages when the legal status of a monitored case changes.

19.2.3   Competitor Intelligence Based on Bibliographic Patent Data

As the patent literature is the most timely and most comprehensive carrier of technological information, it is only logical to recommend the exploitation of patent family and legal status data banks for *competitor intelligence* [19.19]–[19.25]. It is certainly not difficult to use bibliographic patent data to compile statistical charts displaying, for example, the development of patent activities in certain areas of technology. Patent offices and other administrative bodies have collected and published such statistical data on a national or regional basis. They are meant for supporting decision-making in economic policy or long-range research and development planning.

Much more relevant to industrial companies, but also much more demanding, is the design of a computerized warning system that alerts specialists at an early stage to new trends in technology from competitors' recent patent activities. Two approaches can be taken:

1) monitoring new developments in specified areas that attract attention due to the sudden appearance of clusters of patent documents and
2) recognition of changes in the research and development policy of specified competitors that become apparent due to unexpected significant patent activities.

A patent family data bank can, in principle, provide this kind of insight if its entries are comprehensive and reliable. Its speedy update makes the information available at an early date, although the data are not published earlier than 18 months after application, and the corresponding research work might have started years before.

In practice, such a project has its subtle problems: the specific areas of technology to be monitored must be suitably defined and expressed in *retrievable* terms. If they are defined too broadly, the clusters sought for get lost in "noise", if they are defined too narrowly, statistical variance raises many false alarms. The International Patent Classification (Section 19.3.3) is not well suited for this purpose, and other classifications (e.g., those allocated by commercial information services) cannot always be used. Fine-tuning of the "thresholds of stimuli", arrived at after extensive experimentation, is necessary.

Patent statistics treat all patent data alike, irrespective of the contents of the patent documents. Therefore, statistical evaluation of family data banks can only be one component of a patent warning system that must be completed by thorough intellectual analysis and assessment of the inventions disclosed.

19.2.4   Patent Citation Networks

In Section 2.3.1 reference is made to the Science Citation Index, and in Section 1.4 to the theory of citation analysis. Generally, there is no reason why citation indexing should not be as useful for patent literature as it is for scientific literature. Nevertheless, although patent citation indexing has been a method practiced in industrial environments and in patent offices on a private file basis, a patent citation file with a wide coverage of patent systems is currently not available; only a few commercial or public patent citation files, limited to one or few patent systems each, are searchable online (Sections 20.6 and 21.3). The problem of patent citation indexing rests in the choice of the citations to be included: It is relatively straightforward to index the prior art cited on the title pages of patent documents under INID Code number 56 (Section 16.2). There is a WIPO standard for citations [19.26], although patent offices do not always comply with it.

Not only patent literature is cited as prior art on title pages; scientific literature, reports, and other nonpatent literature are also cited and should be included in citation indexing.

Quite often, however, prior art is also cited by the applicant in the introductory part of the description of the invention (Section 16.3). These citations are sometimes scattered throughout the text and are not easily located, but they are highly relevant.

The matter is further complicated by the fact that citations pertinent to the assessment of an invention are distributed over several documents. Unexamined applications only contain citations under INID Code number 56 if the search report (requested or obligatory) was available before the publication date. Many European search reports are issued as separate documents. As long as an application has not been subjected to examination, the citations are preliminary in character. Citations considered by the examiner are published in the corresponding patent specifications.

Much more difficult to locate are citations that come up after grant, e.g., in opposition procedures, in nullity suits, or infringement litigations. Inclusion of these citations poses a difficult task for data base producers.

The collected citations have to be allocated to the respective patent families, both as "warrants of apprehension" of the inventions and as tools for finding related prior art via the citation networks.

Citation files must be maintained as *living files* because it might take years after the publication of the first case (Section 19.2.1) until the the last entries are captured.

Establishing and updating comprehensive patent citation files require considerable effort, and the major commercial patent information services have been fairly reluctant to users' requests for patent citation data bases with a wide coverage.

19.3   Indexing and Retrieval of the Technical Disclosure of Patent Literature

19.3.1   Introduction

The objective of indexing is to compress the information contained in a document into a sufficiently small number of controlled descriptors which can be more easily stored and more reliably retrieved than the full text. This can be done in a variety of ways. *Consistency* is an important aspect of indexing.

The methods used for indexing patent literature are basically the same as for other technical literature. Only approaches that are characteristic of indexing (chemical) patent literature are discussed here.

Representation of Information. The technical disclosure of patent literature is expressed either verbally, in numerical data, in chemical structural formulae, or drawings; each mode of expression has its significance and problems.

Verbal Information. A great deal of scientific and technical chemical information can only be expressed in verbal form. In certain areas of chemistry (e.g., polymer chemistry) verbal descriptions predominate. For reliable indexing, technical terms used in patent documents should be well-defined standard terms, but this is not always the case. When novel concepts are to be described for which standardized terminology is not available, an applicant must coin new terms of his own, which may be expressed differently by another applicant. Although the patent offices encourage applicants to use commonly agreed terminology, this is limited by the freedom an applicant must be granted for describing his invention properly. On the other hand, there is a tendency in patent literature to express deliberately technical concepts in vague and general terms (Section 12.1). It is the indexer's responsibility to select the relevant passages of the disclosure, to identify the underlying concepts in context, and translate these consistently into the indexing language. To be able to do this, he must be knowledgeable in the respective area of technology.

Numerical Data. Numerical data, although apparently the most "refined" form of technical information, are less useful as search criteria in patent searches, for three main reasons:

1) In patent documents, quantifiable data will most often be expressed as ranges, rather than in single numbers; data ranges are more difficult and less reliable to search in.
2) In different countries, numerical values might be expressed according to different national standards; translating values from one standard into another is problematic and error-prone.
3) When no agreement upon standard exists, as is often the case for novel inventions, the applicant will define numerical values ad hoc for the purposes of his particular invention; these data are not comparable with others and cannot be chosen as search criteria.

Thus, when a search is expected to be complete, numerical data should not be used in the query.

Structural Formulae. Structural formulae are the most significant medium of chemical reasoning and communication. Methods for indexing the information contained in structural formulae have always been the focus of chemical information management. Chemical structure is dealt with from the point of view of scientific literature in Chapter 10, so that only some additional remarks on structural formulae in *patent literature* are required here.

Chemical structures published in scientific literature can now be searched online in a wide variety of commercial data bases, but few commercial data bases cover chemical patent literature comprehensively. Among these, recall-type data bases (Section 19.1) are rare, and until recently, their structure input was based on fragmentation coding (Section 10.1.2), rather than on connection tables (Section 10.1.4) because the indexing and retrieval problems of generic chemical structure handling had not yet been overcome. Structure indexing and retrieval of patent literature are currently in a phase of fundamental change, the bulk of retrievable structure information can still only be searched by traditional techniques (classifications, fragmentation codes).

Graphical Information. Graphical information other than structural formulae (e.g., schemes, flowsheets, and drawings which are indispensable in chemical engineering and other process-oriented sectors of chemistry) have remained largely inaccessible to direct indexing. Indexing of and searching by textual information as well as visual-intellectual checking of the graphical parts of the retrieved documents is still common practice. Fast, high-resolution display methods (Section 8.2.9) are important for enhancing the efficiency of visual-intellectual retrieval.

Relevance. Indexing is preceded by selecting words and statements containing relevant information from the full text of the patent document, i.e., the indexer has to make a decision concerning *relevance*. The notion of relevance has been discussed in the literature for years, many attempts have been made to measure it reproducibly [19.27]–[19.31]. Relevance decisions must be taken by different people involved in the successive steps of the information process (abstractors, indexers, searchers, users). It appears that a relevance decision very much depends on the state of knowledge and on the experience of the person who has to take it [19.32], and is, to a considerable extent, subjective. Moreover, a relevance decision is always influenced by a specific situation influenced by the times; nobody can safely predict which details of a disclosure might become particularly important in future searches. In our electronic age, information that has not been selected for indexing is virtually lost forever. Searchers should bear this in mind when asked to perform comprehensive searches.

Context. Compression by indexing makes information more amenable to retrieval but leads to loss of context. Understanding the meaning of an expression is greatly improved by its context, this is particularly so for the inherently complex concepts encountered in patent literature. Context in its richest form is offered in *interpersonal discourse*; the interplay of question and answer is the most effective kind of information exchange. *Full texts* (Section 19.3.5) also present words in their contexts. In *abstracts* (Section 19.3.2), context is reduced as compared with the corresponding full text; the quality of an abstract heavily depends on the preserved context. *Classifications* (Section 19.3.3) still preserve rudimentary context. *Coordinate indexing* (Sections 19.3.4 and 19.3.5) dissolves context entirely. Post-coordination of descriptors by Boolean and other operators (Section 8.4.4) hardly compensates for the loss of context, and inexorably confronts the searcher with the recall-precision dilemma (Section 19.1).

Search Strategy. Patent search strategies must be carefully adapted to the users' real requirements which have to be identified by painstaking inquiry. It might appear trivial to say this as user satisfaction has always been the information manager's overriding objective. However, it should be realized that patent searches can be time-consuming and expensive, and it is a big mistake to waste scarce resources by obtaining meaningless or even misleading search results. User requirements with respect to searches in the patent literature can be grouped into three main categories: informative searches, prior art searches, and infringement searches.

Informative Search. Informative searches, mostly asked for by research and development staff, are expected to be quick and precise, not exhaustive, and not "polluted" with lots of irrelevant citations, and can sometimes be limited to one or two data bases or to time ranges. Depending on the patent data base searched in, the percentage of irrelevant hits might nevertheless be so high, that the user would welcome having the relevant hits sifted by the searcher.

In informative searches, patent literature is generally treated as scientific literature. Simple structure searches in suitable data bases can and should be performed by *end-users* (Chap. 20.8).

Prior Art Search. Prior art and infringement searches are directly connected with patent matters, they are mostly requested by patent specialists and carried out by professional searchers. A prior art search usually starts from a patent claim arising from the claim of a competitor's patent application or patent to be invalidated in an opposition procedure or nullity suit. Another common reason for prior art search is a drafted claim of a corporate patent application that is to be filed and for which possible objections of the patent examiner must be anticipated. The objective of a prior art search is to unearth any documents (not only patent documents) that describe the invention claimed or a *specific embodiment of it*. If the claimed invention is imagined as occupying a position in a hierarchical arrangement of disclosures, the most general description at the top and the most specific at the bottom, then every description is prior art which is on the same level as the claimed invention or on a lower level. It is for this reason that prior art searches are sometimes called *down-searches*.

The most gratifying outcome of a prior art search (from the viewpoint of the searcher, not necessarily from the viewpoint of the user who asked for the search) is to detect an identical anticipation of the claimed invention. Direct hits are not very frequent, however. Therefore, the searcher should also concentrate his attention and strategy on descriptions close to the target invention to prove its obviousness. It is difficult to determine when to stop searching if no serviceable prior art is found. This must be decided on the basis of the importance of the case: if much is at stake, the search for prior art might have to continue until the last available source of information is exhausted. In principle, there is no limit as to publication date, language or media of divulgation. In practice, time and costs always set a limit.

Infringement Search. The object of an infringement search is to identify every existing patent or pending patent application that might impede the sale of a product awaiting launch or its manufacture. Patent infringement must be avoided at all costs, hence infringement searches are the most demanding patent searches.

Infringement searching is a three-step process: In the first step "suspicious" patent documents are identified, in the second and third steps their patent family and legal status are verified. An infringement search starts from the product or process in question and is directed to retrieve every document that contains a patent claim covering the product or process as such or a *general representation of it* (e.g., all classes of chemical compounds embracing the target product). In analogy to down-searches, infringement searches are sometimes called *up-searches*.

If the product is a mixture of chemical compounds, infringement searches must take every nontrivial component of the mixture into account.

Infringement searching will be limited to those countries in which the product is to be marketed or the process is to be employed. In the second step of an infringement search, recourse is therefore made to the corresponding patent families to identify the patents or patent applications filed in these countries. In a third step, their legal status is checked. Only those patents that are still in force and those patent applications that are still pending are relevant for infringement investigations. Expired patents and abandoned patent applications need not be considered.

19.3.2 Patent Abstracts

Abstracts play an important role in processing patent information. They form the intermediate stage between the full texts of the bulky patent documents and their computer-searcheable representations, exhibiting to the reader the essence of the disclosures in context. To be useful, patent abstracts must be prepared by specialists who should be able to present complex technological information accurately and in a form that is quickly digestible in browsing [19.33], [19.34].

The abstracts printed on the title pages of patent documents (Section 16.2, INID Code number 57) seldom fulfill these stringent requirements because they have no legal significance, and patent offices do not put much emphasis on controlling their quality.

High-quality patent abstracts have been available for many years from a commercial information service, *Derwent Publications* (Fig. 16). They serve a number of purposes in patent information management: Packaged suitably and distributed according to the needs of the readers they are welcome media providing up-to-date information on new technical developments. They may be grouped according to the IPC or other classification symbols shown in the abstract headings and used directly by the addressees as card files for manual search purposes.

Patent searchers use abstracts, displayed on their personal computer monitor, to assess the relevance of search results obtained in electronic retrieval systems and will modify their search strategies accordingly. The selected patent abstracts can subsequently be sent to the users.

Unfortunately, the abstracts presently offered for monitor display are not of the same quality as those shown in Figure 16. These are available either printed on paper (single abstracts on cardboard or bound in bulletins) or on microforms. It is expected that high-quality patent abstracts will soon be available in digitized form on magnetic tape or optical storage media [19.35]. This will offer novel possibilities of supplying current awareness information or search results to users via communication networks.

19.3.3 The International Patent Classification (IPC) and Other Classifications

A *classification* is a list of concepts taken from a certain field of human knowledge, and appropriate descriptors allocated to them. It has the following features:

1) The concepts are mutually exclusive, i.e., each concept however complex occupies one and only one position in the list and hence is allocated one and only one descriptor; if a (new) concept does not fit in, an "others" descriptor is allocated to it.
2) The list of descriptors is fixed, no new descriptors can be added as long as the particular version of the classification is in use.
3) Allocation of descriptors to multifacet concepts is governed by *preference rules* indicating which facet determines the allocation.
4) The concepts are arranged in hierarchical order.

| 88-221149/32   A88 D15 J01 (A14 A26)   AGEN 22.01.87 | A(10-E21B, 11-B5A, 12-W11A) D(4-A1A, 4-A1E, 4-B6) J(1-C3) D0040 |

AGENCY OF IND SCI TECH   *DE 3801-690-A*
03.07.87-JP-165285 (+JP-011337) *(04.08.88)* B01d-13 B01d-53/22
C02f-01/44 C08j-03/24 C08j-05/22
Poly-ion complex polymer membrane - for sepg. water from organic
substance in liq. or vapour form
    C88-098649

(THF, dioxane), organic acids (formic, acetic), aldehydes
(acetaldehyde, propionaldehyde) or amines (pyridine,
picoline) or gaseous mixts. contg. water and one or more
of these cpds.

Other Priority : 04.03.87-JP-047495 03.07.87-JP-165284

PREFERRED COMPOSITION
    (I) is crosslinked with a crosslinking agent (pref. a
polyfunctional epoxide, amine, methylolmelamine or NCO
cpd.) forming a covalent bond with the polymer and is
polyacrylic acid or its metal or ammonium salt.
    The cationic gp. of (II) is a prim., sec., tert. or quat.
amino gp. and (II)   pref. is polyallylamine or polyethylene-
imine; or (II) has a quat. ammonium salt gp. in its main
chain and pref. is of formula (IIA), (IIB), (IIC) or (IID):

Polymer membrane for sepg. water from an organic substance
by pervaporation of permeation of water vapour consists of
a synthetic polymer (I) with an anionic gp. associated with a
synthetic polymer (II) with a cationic gp. by an ionic bond to
form a polyion complex on the surface of the membrane
and/or in the membrane.

USE/ADVANTAGE
    The membrane is water-resistant and durable and has a
satisfactory permeation rate and high sepn. coefft. over a
wide concn. range of an organic substance in soln. It is
useful for sepg. water from e.g. alcohols (MeOH, EtOH,
1-PrOH, 2-PrOH, n-BuOH), ketones (acetone, MEK, ethers

$$\left[\begin{array}{c} R_3 \;\; X^{\ominus} \\ | \\ N - R_1 - \\ | \\ R_4 \end{array} \begin{array}{c} R_5 \;\; X^{\ominus} \\ | \\ N^{\oplus} - R_2 \\ | \\ R_6 \end{array} \right]_n \quad (IIA)$$

DE3801690-A+

$$\left[\begin{array}{c} CH_3 \;\; X^{\ominus} \\ | \\ N^{\oplus} - CH_2 - CH_2 - \\ | \\ CH_3 \end{array} \begin{array}{c} CH_3 \;\; X^{\ominus} \\ | \\ N^{\oplus} - CH_2 - \bigcirc - CH_2 \\ | \\ CH_3 \end{array} \right]_n$$

(IIB)

$$\left[\begin{array}{c} CH_3 \;\; X^{\ominus} \\ | \\ N^{\oplus} - CH_2 - CH_2 - CH_2 \\ | \\ CH_3 \end{array} \right]_n \quad (IIC)$$

$$\left[\begin{array}{c} CH_3 \;\; X^{\ominus} \\ | \\ N^{\oplus} - CH_2 - CH - CH_2 \\ | \qquad\qquad | \\ CH_3 \qquad\quad OH \end{array} \right]_n \quad (IID)$$

R_1 and R_2 = alkylene with $\geqslant$ 2C or a hydroxyalkylene,
                alicyclic or aromatic gp.;
R_{3-6} = 1-3C (hydroxy)alkyl;
X^- = a halide counter-ion.
    The membrane pref. is a composite membrane with a skin
of the polyion complex and a porous substrate.

PREPARATION
    The membrane can be prepd. by dipping a (I) or (II)
membrane, which has been rendered insol. by crosslinking,
in a (II) or (I) soln. For a composite membrane, a porous
polymer membrane with an insolubilised (I) coating is dipped
in a (II) soln.

EXAMPLE
    An aq. soln. of polyacrylic acid (viscosity 24 cP for 1%
aq. soln.) was neutralised with aq. NH_3 and diluted with
water to give a 0.5% polyammonium acrylate soln. This was
spread on a polyethersulphone ultrafiltration membrane and
dried in air to give a 0.4 μ thick soln. The composite
membrane was dipped in a slightly acidic 0.5% soln. of
polyallylamine hydrochloride (pH 3.5) in 1 : 1 EtOH/water
for 10 min. at room temp., giving a polyion complex
membrane.

DE3801690-A+/1

| 88-221149/32 | A 95 : 5 EtOH/water mixt. at 0.1 | | D0041 |

kg/cm² excess pressure and 70°C was
supplied to the prim. (main) side of (A) the complex
membrane or (B) the original polyammonium acrylate
membrane. The permeation rate was (A) 0.69, (B) 0.54
kg/m².h; and the sepn. coefft. (A) 107, (B) 38.
(22pp016RBHDwgNo0/0).

DE3801690-A/2

Figure 16. A patent abstract as published by Derwent Publications Ltd, London
Reprinted by permission of the publishers.

The advantage of classifications over other indexing schemes is their unambiguousness. Documents can be filed in a linear arrangement mirroring the classification used; only one position has to be looked up when a specific concept is searched for, and all the relevant documents can be expected to be assembled there.

If the list of descriptors is a hierarchical notation, the scope of a search can easily be broadened (generalized) or narrowed (specialized) if too few or too many documents appear to be retrieved.

The advantages of a classification can, however, only be exploited if indexing is consistent and in strict compliance with the classification rules.

On the other hand, classifications are affected with severe problems: In multifacet concepts, those facets that do not, according to the preference rules, determine the allocation of the descriptors, are lost and can no longer be searched as such. Relevant documents might be missed.

Since the list of descriptors (*class symbols*) available for indexing is fixed, a classification must be revised from time to time to include new concepts that have emerged in the course of scientific and technological development. It is impossible to predict which positions of a classification might be subject to revision, changes occur at random. This means that each revision creates a new version of the classification which is, at least in part, incompatible with earlier versions. Searches ranging over more than one revision period must be adjusted accordingly.

International Patent Classification [19.36]–[19.38]. The International Patent Classification (IPC) dates back to 1954 when 15 European countries concluded the *European Convention on the International Patent Classification* that came into force for seven countries in 1955. In 1971, the IPC was based on an international convention, the *Strasbourg Agreement Concerning the International Patent Classification*. The WIPO was entrusted with its administration and with organizing the regular revisions taking place every five years. Presently (1991), 26 states are party to the IPC Agreement. The IPC is also applied to the documents issued in the existing supranational patent systems, and it is widely used by patent offices for arranging their search files.

The first edition of the IPC was issued in 1968, the current edition is the fifth (IPC^5), obligatory from 1 January, 1990, comprising nine volumes, dividing the entire field of technology into eight sections and approximately 64 000 units (classes, subclasses, groups, and subgroups) [19.39]. Figure 17 shows a page of IPC^5. Ninety-two classification rules must be complied with when using the IPC. Volume 9 of the IPC Manual gives extensive guidance to users.

The IPC is mainly function-oriented [19.40] and is reasonably well adapted to the needs of engineers. Chemical inventions usually cannot be sufficiently described in terms of function; classes of chemical compounds as well as processes for their preparation and use are not amenable to classification by the IPC. For these and other reasons the IPC has recently been supplemented by *index classes* which can only partly overcome these inherent problems.

Another serious difficulty of the IPC is the lack of consistency of its application in different patent offices and over time [19.41]–[19.43]. Although the WIPO officials try hard, the chances of achieving consistency in the application of the IPC

C 07 D

227/08 . . . Oxygen atoms [2]
227/087 . . . . One doubly-bound oxygen atom in position 2, e.g. lactams [3]
227/093 . . . . Two doubly-bound oxygen atoms attached to the carbon atoms adjacent to the ring nitrogen atom, e.g. dicarboxylic acid imides [3]
227/10 . . . Nitrogen atoms not forming part of a nitro radical [2]
227/12 . with hetero atoms directly attached to the ring nitrogen atom [2]

229/00 Heterocyclic compounds containing rings of less than five members having two nitrogen atoms as the only ring hetero atoms [2]
229/02 . containing three-membered rings [3]

231/00 Heterocyclic compounds containing 1,2-diazole or hydrogenated 1,2-diazole rings [2]
231/02 . not condensed with other rings [2]
231/04 . . having no double bonds between ring members or between ring members and non-ring members [2]
231/06 . . having one double bond between ring members or between a ring member and a non-ring member [2]
231/08 . . . with oxygen or sulfur atoms directly attached to ring carbon atoms [2]
231/10 . . having two or three double bonds between ring members or between ring members and non-ring members [2]
231/12 . . . with only hydrogen atoms, hydrocarbon or substituted hydrocarbon radicals, directly attached to ring carbon atoms [2]
231/14 . . . with hetero atoms or with carbon atoms having three bonds to hetero atoms with at the most one bond to halogen, e.g. ester or nitrile radicals, directly attached to ring carbon atoms [2]
231/16 . . . . Halogen atoms or nitro radicals [2]
231/18 . . . . One oxygen or sulfur atom [2]
231/20 . . . . . One oxygen atom attached in position 3 or 5 [2]
231/22 . . . . . . with aryl radicals attached to ring nitrogen atoms [2]
231/24 . . . . having sulfone or sulfonic acid radicals in the molecule [2]
231/26 . . . . 1-Phenyl-3-methyl-5-pyrazolones, unsubstituted or substituted on the phenyl ring [2]
231/28 . . . . Two oxygen or sulfur atoms [2]
231/30 . . . . . attached in position 3 and 5 [2]
231/32 . . . . . . Oxygen atoms [2]
231/34 . . . . with only hydrogen atoms or radicals containing only hydrogen and carbon atoms, attached in position 4 [2]
231/36 . . . . with hydrocarbon radicals, substituted by hetero atoms, attached in position 4 [2]
231/38 . . . . Nitrogen atoms (nitro radicals 231/16) [2]
231/40 . . . . . Acylated on said nitrogen atom [2]
231/42 . . . . . Benzene-sulfonamido pyrazoles [2]

231/44 . . . . Oxygen and nitrogen or sulfur and nitrogen atoms [2]
231/46 . . . . . Oxygen atom in position 3 or 5 and nitrogen atom in position 4 [2]
231/48 . . . . . . with hydrocarbon radicals attached to said nitrogen atom [2]
231/50 . . . . . . Acylated on said nitrogen atom [2]
231/52 . . . . . Oxygen atom in position 3 and nitrogen atom in position 5, or vice-versa [2]
231/54 . condensed with carbocyclic rings or ring systems [2]
231/56 . . Benzopyrazoles; Hydrogenated benzopyrazoles [2]

233/00 Heterocyclic compounds containing 1,3-diazole or hydrogenated 1,3-diazole rings, not condensed with other rings [2]
233/02 . having no double bonds between ring members or between ring members and non-ring members [2]
233/04 . having one double bond between ring members or between a ring member and a non-ring member [2]
233/06 . . with only hydrogen atoms or radicals containing only hydrogen and carbon atoms, directly attached to ring carbon atoms [2]
233/08 . . . with alkyl radicals, containing more than four carbon atoms, directly attached to ring carbon atoms [2]
233/10 . . . with only hydrogen atoms or radicals containing only hydrogen and carbon atoms, directly attached to ring nitrogen atoms [2]
233/12 . . . with substituted hydrocarbon radicals attached to ring nitrogen atoms [2]
233/14 . . . . Radicals substituted by oxygen atoms [2]
233/16 . . . . Radicals substituted by nitrogen atoms [2]
233/18 . . . . Radicals substituted by carbon atoms having three bonds to hetero atoms with at the most one bond to halogen, e.g. ester or nitrile radicals [2]
233/20 . . with substituted hydrocarbon radicals, directly attached to ring carbon atoms [2]
233/22 . . . Radicals substituted by oxygen atoms [2]
233/24 . . . Radicals substituted by nitrogen atoms not forming part of a nitro radical [2]
233/26 . . . Radicals substituted by carbon atoms having three bonds to hetero atoms [2]
233/28 . . with hetero atoms or with carbon atoms having three bonds to hetero atoms with at the most one bond to halogen, e.g. ester or nitrile radicals, directly attached to ring carbon atoms [2]
233/30 . . . Oxygen or sulfur atoms [2]
233/32 . . . . One oxygen atom [2]
233/34 . . . . . Ethylene-urea [2]
233/36 . . . . . with hydrocarbon radicals, substituted by nitrogen atoms, attached to ring nitrogen atoms [2]
233/38 . . . . . with acyl radicals or hetero atoms directly attached to ring nitrogen atoms [2]
233/40 . . . . Two or more oxygen atoms [2]

Figure 17. A page taken from the IPC[5]

among a vast number of classifiers employed in so many patent offices are small. Readers can easily detect variances when spotchecking the IPC classes allocated to the different documents of a patent family.

In patent systems with deferred examination and early publication (Section 12.2), the IPC symbols printed on the title pages of published patent applications are obviously preliminary; many applications are reclassified in the course of substantial examination. This adds another element of uncertainty to the use of the IPC.

The lack of consistency of IPC classes for incoming patent documents has given rise to the practice used by the European Patent Office of complete reclassification of all documents entering the EPO's search files.

Summing up, it seems fair to say that IPC classes printed on the title pages of patent documents do not lend themselves to performing reliable patent searches in chemistry, although they may be useful in searches that do not have to be comprehensive.

Other Classifications. Some patent offices still apply national classifications in addition to the IPC. These are based on different philosophies from that of the IPC, and hence cannot be translated into it in toto. The USPTO, for example, still classifies according to the *US national classification*, the national classes subsequently being machine-translated into IPC classes; incorrect IPC symbols are not infrequent on U.S. patent documents.

19.3.4   Indexing and Retrieval of Chemical Structures

Chemical structure representation and structure searches have already been dealt with from the viewpoint of scientific literature in Chapter 10. This section can therefore be confined to specific aspects of structure handling in patent information management.

In the management of scientific chemical literature, topological input and retrieval methods based on the connection table concept (Section 10.1.4) have won the field. The reasons are obvious. Topological methods make full use of the structural formula language that chemists have been trained to think and communicate in. Any substructure is preserved in a connection table and can be retrieved. The indexing step, in which the structural formula is translated into its connection table, can be entrusted to computer software.

In contrast, a fragmentation code (Section 10.1.2), inexorably complex when designed to approximate topological methods in performance, must be learned and memorized as a separate language just for the purpose of input and retrieval. Substructure searches are limited to preformed fragments, and structure input is an intellectually demanding procedure that can only be done by trained staff. However, the advantages of topological methods can only be exploited fully in the handling of formulae of *individual compounds*. This is so because connection tables map the connections (bonds) between atoms; when a structural formula contains variable attachments or substructures, as do generic formulae, then a number of problems arise [19.44]–[19.50].

A fragmentation code is a list of preformed structure fragments (substructures) and of descriptors allocated to each fragment. To be indexed, a formula is dissected into the preformed fragments which in turn are indexed by their descriptors. The precision with which a structure can be represented in a fragmentation code depends on the size of the descriptor list (fragment vocabulary) and on the degree of accuracy with which the connections between the fragments are indexable (syntax) [19.51].

Topological methods are at their best when *individual compounds* are to be indexed; in contrast, fragment codes are very suitable for indexing *classes of compounds*. Claims of chemical patent documents very often pertain to classes of chemical compounds, rather than to individual compounds, and the classes are typically expressed in terms of generic formulae. It is for this reason that fragment codes still play an important role in indexing and retrieval of chemical patent literature. Very powerful indexing and retrieval systems based on fragmentation coding have been developed and have been in use for many years (Sections 20.3 and 20.4).

Anybody developing and maintaining a fragmentation code system will sooner or later be confronted with a dilemma: In view of the steadily increasing amount of chemical patent literature issued annually, he is, on the one hand, forced to enhance continuously the performance (resolution, particularly with respect to syntax) of his fragmentation code in order to keep its precision; on the other hand, increasing sophistication of the system imposes a growing intellectual burden on the indexers, rendering the system more error-prone, and makes the employment of additional trained personnel inexorable. As staff salaries are the major cost elements of fragmentation code systems, the costs rise in proportion to the number of patent documents processed.

These problems are much less felt in topological systems. The indexing procedure is carried out by computer software, input work is mainly manual and can be done by a workforce with less training. No adjustment of indexing precision is necessary because connection tables preserve the entire structural information. These circumstances have for a long time spurred endeavors to make wider use of topological methods in chemical patent information management, and, in particular, to extend them to the handling of generic formulae.

Two types of generic formulae are encountered in patent literature. A *type 1 generic formula* (Fig. 18) represents a finite, albeit sometimes extensive, class of chemical compounds. The individual compounds covered can be enumerated by permutation of the variable substituents or attachments in the formula. Topological methods were successfully applied to the indexing of type 1 generic formulae as early as in the late 1950s in the chemical industry [19.46], [19.52].

A type 2 generic formula (Fig. 13; page 167) represents virtually an *infinite* class of chemical compounds. The individual compounds covered cannot be enumerated by permutation because the substituents represent substructure classes of infinite extension (e.g., alkyl, aryl, heterocyclyl) and are sometimes interpretable only with recourse to the context in the description. Here lies the frontier of current research in generic chemical structure handling (Section 10.4.1) [19.53]–[19.55].

Generic formulae pose problems not only in indexing but also in retrieval [19.56], [19.57]. Extensive generic classes must be searchable with acceptable speed for all individual compounds covered that might be considered as prior art. Equally, infringement searches must be feasible in these classes. Although progress in develop-

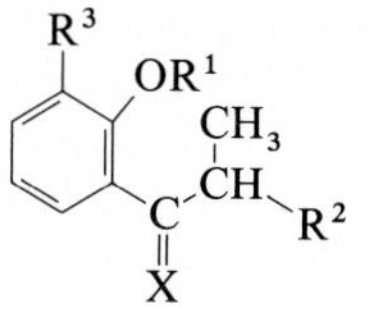

$R^1 = H, CH_3, C_2H_5$

$R^2 = H, CH_3$

$R^3 = H, -CH\begin{smallmatrix}CH_3\\CH_3\end{smallmatrix}$

$X = O, S$

Figure 18. A type 1 generic formula

ing such retrieval systems has been very promising, it will take some time until full-fledged structure and substructure searching becomes a convenient and reliable option in large topological files covering generic formulae.

Extremely broad patent claims based on generic formulae that encompass very large numbers of chemical compounds have been appearing increasingly in patent documents and have given rise to controversial discussions [19.58]–[19.62]. Also, the question of novelty of chemical compounds covered by broad generic classes, but not explicitly described, has been raised. Patent offices and courts have dealt with this problem, but opinions still differ. Patent information managers should watch the outcome closely.

19.3.5   Indexing and Retrieval of Non-Structural Information

Chemical patent documents always contain a considerable proportion of *verbal* information which represents an indispensable complement to structural information: reaction conditions, process details, shapes and uses of products etc. must be described in words. These verbal concepts are, unfortunately, not always expressed in standard terminology (Section 19.3.1). In fact, technical concepts can be set forth in many different ways, ranging from single words to clauses or even paragraphs, all with the same meaning: all these descriptions are *synonymous*. Even the most experienced specialist is unable to recall all and every possible variant of describing or circumscribing a given technical concept [19.11]. Complete search results can only be expected when the vocabulary used for indexing and retrieval is kept in strict control.

There are two different approaches of vocabulary control: the fixed vocabulary and the thesaurus methods [19.63].

Fixed Vocabulary. Fixed (preformed) vocabularies must be revised from time to time in the course of technical development. In coordinate indexing vocabularies (see below) this is not as cumbersome as in classifications (Section 19.3.3); it is useful to register the date on which a descriptor is added to the vocabulary.

Thesaurus [19.64]–[19.66]. The vocabulary of a thesaurus is not preformed. Instead, it is built up stepwise by collecting the descriptors used in indexing which are then arranged and listed in hierarchical order (thesaurus) and put to the searchers' disposal. In this way, the vocabulary is continuously adapted to recent technology.

Whichever method of vocabulary control is employed, the descriptors are used for *coordinate indexing*: each descriptor stands for a specific facet of a technical concept (not for an entire complex concept as in classifications). The indexer chooses as many descriptors as he feels are necessary to represent the technical contents of a patent document, the descriptors being allocated independently of each other. Coordination of the descriptors is done at the retrieval stage (*post-coordination*) by means of Boolean and other operators. As the context of the disclosure disappears in coordinate indexing (Section 19.3.1), considerable amounts of irrelevant hits might turn up in post-coordinated searches. When complete search results are expected, this must be tolerated.

Indexing verbal patent information is a demanding task. Expert systems have been used or proposed for easing the indexers' burden [19.67].

Full-Text Retrieval. The most radical approach to overcome indexing problems is not to index but to subject the full texts of patent documents to computer search. The difficulties then encountered are tremendous. Besides the problem of (timely) obtaining machine-readable texts of patent documents issued by a great number of patent offices, these texts are written in different languages, each requiring special sophisticated computer software for its analysis. Available language processing software is still being developed and can not adequately cope with the complex technical disclosure in patent literature. Reports on full-text retrieval experiments carried out in considerably less complex material show distressingly low recall [19.68]–[19.71], totally inacceptable to patent information management.

This does not mean that patent information is entirely unsuitable for text searches. In combination with the retrieval of indexed information, *text searching in patent abstracts* (written in a unique language) is quite useful when high precision (and not high recall) is aimed at [19.72].

Patent offices have established, or are planning to establish, *text display systems* for their examiners using the IPC as access tool but also allowing text searching [19.73]–[19.74].

20 Existing Abstracting and Indexing Services Covering Patent Literature

20.1 Introduction

The December 1990 issue of the WIPO Handbook of Industrial Property Information and Documentation [20.1] lists 35 commercial or public data bases consisting exclusively or to a large extent of references to patent documents. They are being offered on 17 different hosts. Two new data bases, Derwent's WPIM and CAS's MARPAT, both specialized in generic structural formulae, must be added to the list. Only one of these 37 data bases does not cover any chemistry. In 18 of the remaining 36 data bases indexed chemical information can be searched, and one contains full texts. Six of the 19 data bases cover the documents of more than five patent systems. The other 17 are data banks containing bibliographic patent data only, five of them cover more than five patent systems.

The six data bases with wide patent system coverage holding indexed chemical information are produced by just three organizations: Chemical Abstracts Service, Derwent Publications, and the EPO, the latter providing in its EDOC data base (offered by Télésystèmes) technical information indexed by the EPO's special version of the IPC (Section 19.3.3). EDOC started in 1969. Chemical patent information retrieval mostly concentrates on the CAS and Derwent data bases (Sections 20.2 and 20.3).

The same situation holds for bibliographic patent data: the five data banks with wide patent system coverage are all produced by INPADOC, which was integrated into the EPO in January 1991 (Section 20.5).

Growing markets and increasing internationalization of business require comprehensive data bases and data banks for monitoring patent activities on a world-wide basis. It follows that retrieval of technical information and patent data concentrates on the data bases and data banks with wide coverage of patent systems and areas of technology. There is obviously very little choice. The other specialized data bases and data banks can be and are used for supplementary searches, particularly when complete search results are expected. Their use is limited by the fact that for accessing them searchers have to learn and memorize additional query and command languages.

Up to the early 1970s chemical companies maintained their own card files or data bases for information retrieval; this was very expensive. Cooperation schemes developed (Section 22.2) which, with few exceptions (Section 20.4), were replaced by the extensive use of commercial and public data bases which at that time had matured into reliable, cost-effective services.

Producing comprehensive chemical patent data bases, holding them at public disposal on big computers, and providing versatile retrieval software for their efficient use requires high financial investment. So, it is only logical that these activities are concentrated in few powerful organizations serving the rest of the world. As a result, however, industrial companies become increasingly dependent on the few existing information services for their patent information management. The only reasonable solution to this is close, trustful cooperation between the users and producers of information in user groups and by exchange of experience to their mutual benefit (Section 22.2.2).

Searches in online data bases have become common practice. About 1000 publications on online searching appear annually. For reasons already explained the utilization of data bases containing scientific and technical information varies considerably, and the trend towards mergers and takeovers of service organizations continues. Activities of individual countries, motivated by national pride and policy, do not make much sense in the international arena of patent information management.

CD-ROM. Producers have reacted to the continuous success of data bases on CD-ROM, but it seems unlikely that CD-ROM data bases will become a severe threat to online patent patent information retrieval: comprehensive patent data bases are too big to be conveniently searched on CD-ROM. Small, specialized online data bases might be replaced by CD-ROM, and well-designed subsets of large data bases offered on CD-ROM, tailored to the needs of specialized user communities, will have a market.

Downloading [20.2], [20.3]. Downloading information from the hosts' mainframes to smaller in-house computers for later use has also been practiced for some time. Two problems have to be faced: Downloading only makes sense if the data are at least as conveniently searchable in the downloaded file as on the host computer. This can be achieved if the data received from the host in answer format are rearranged by powerful data base management software into a user-friendly search format. Even then, the versatility of modern retrieval software provided by the hosts is not easy to beat. Moreover, the downloaded file must be frequently updated to keep up in currency with the mainframe data base. In these circumstances, downloading requires some financial effort and is cost-effective only when the downloaded file is frequently searched in.

20.2   Chemical Abstracts Service

The data bases provided by Chemical Abstract Service (CAS) have already been dealt with extensively in Section 2.2; some additional remarks remain to be added from the point of view of patent information management.

CASONLINE. The highly successful chemical information system CASONLINE has been geared to scientific literature from its start, hence patent documents have been treated as a special kind of scientific literature. The abstracts, the selection of chemical compounds and verbal concepts for indexing [20.4] and the CA Registry System for individual compounds rely heavily on facts disclosed in a fashion complying with the standards of scientific accuracy. This philosophy has made CASONLINE a precision-type system, well suited for searches for structures and substructures of individual compounds described in patent documents. CASONLINE captures the hard core of the disclosure; in this respect, CASONLINE is an ideal complement to Derwent's WPI files (Section 20.3). The scope of the patent claims cannot readily be inferred, neither from the abstracts nor from the indexed structures and descriptors.

Characterization of CASONLINE as a precision-type system does not mean that high-recall searches are not possible; they are possible but require a great deal of experience and familiarity with the subtleties of the CASONLINE files which usually only professional searchers acquire.

MARPAT. The imperfections of the CASONLINE system with respect to chemical patent information have been felt by Chemical Abstracts Service for some time. In 1990, CAS introduced a new data base called MARPAT which contains searchable representations of generic chemical structures from the patent literature. MARPAT is a topological, document-based file that is closely linked to the files of CASONLINE and supplements this system. MARPAT covers chemical patent literature from 1988 on for all the patent systems considered in CASONLINE, except the Soviet Union. Access is made to MARPAT via the CA abstract number. The search logic appears well designed; the interplay of the different levels of generalization, a difficult problem in generics retrieval, seems to meet the requirements of the users. Interpretation of the search results is demanding and requires skill and attention; MARPAT is designed for the professional searcher. The development of MARPAT is not yet finished, further improvements are possible and will be implemented in due course [20.5]–[20.6].

20.3   **Derwent Publications** [20.7]–[20.10]

Among the commercial patent information services, Derwent Publications holds an eminent place. Derwent offers a comprehensive range of services giving maximum support to industrial patent information management: printed information services, online data bases, abstracts, patent documents supply, customized current awareness bulletins, statistical patent data analysis, bureau searches etc. The Derwent services cover all sectors of technology and the documents of 31 patent systems, in part from 1963 and full-scale from 1970 on (with minor variations) with two additional defensive publication bulletins. Patent family data are accessible online.

Printed Products. Derwent produces two types of abstracts: *Alerting Abstracts* and *Documentation Abstracts*, the former being terser and more current; the latter, based on first cases (Chap. 19.2.1), are high-quality abstracts displaying bibliographic data and giving detailed account of the invention (Fig. 16; page 188). The abstracts are claim-oriented, with additional information on examples, uses, etc. The Derwent Documentation Abstracts (available for chemistry only) are widely recognized as the best patent abstracts in the market-place. The abstracts can be purchased on cardboard (single page), bound in bulletins, on microforms and, in the near future, in digitized form.

There are four separate services:

1) *World Patent Abstracts* (WPA): patent abstracts by country and by subject, the latter also covering general and mechanical technology,
2) *Chemical Patents Index* (CPI): abstracts on 12 sections of chemical technology covering 7000 patents weekly, a corresponding data base, profile bulletins, and other supplementary products,
3) *Electrical Patents Index* (EPI), and
4) *WPI Gazette Service*.

The 12 sections of the Chemical Patents Index are:

– General Chemicals,
– Plastics and Polymers,
– Pharmaceuticals,
– Agricultural Chemicals,
– Food and Detergents,
– Textiles and Paper,
– Petroleum,
– Printing, Coating and Photographic,
– Chemical Engineering,
– Nucleonics, Explosives, Protection,
– Metallurgy,
– Glass, Refractories, and Electrochemistry.

For each of these sections, Alerting Bulletins (country or classified order, the CPI Classification comprising 135 classes) and Documentation Abstracts are issued. *Standard* and *customized bulletins* are available for a wide variety of areas.

Online Data bases. The *WPI data base* is offered online by three major host organizations: Maxwell Online, Dialog Information Services, and Télésystèmes-Questel. It is also loaded in Tokyo by SDC Japan. The WPI is divided into two files: WPI (1963 to 1980) and WPIL (1981 to date). Chemical information is retrievable by Derwent Classification, by indexed descriptors (all Sections in the Manual Code, four Sections additionally in deeper coordinate-indexed codes, now switching into topological coding), by title and abstracts words, and numerous bibliographic parameters.

WPI and WPIL are recall-type files. The wide variety of available search parameters and the familiarity required for formulating efficient search strategies make both files unsuitable for nonspecialist searching.

Recently, Derwent has launched the file *WPIM* for structure retrieval in generic formulae [19.50]. The corresponding sophisticated software has been developed in cooperation with Télésystèmes-Questel and the French Patent Office. It appears to be based on a philosophy different from that of CAS's MARPAT, and its refinement is still under way.

For the statistical analysis of bibliographic patent data (Section 19.2.3) Derwent has been offering a software package, *PATSTAT*, which can be used either online or in downloaded Derwent data.

20.4   International Documentation in Chemistry (IDC)

IDC and its GREMAS code are briefly mentioned in Section 10.1.2. As the IDC services are available only to its share-holding chemical companies, the IDC system will be dealt with here mainly for historical and methodological reasons.

GREMAS, a high-performance fragmentation code system, goes back to work done at Hoechst AG by ROBERT FUGMANN et al. [20.11], starting in the late 1950s. The system has been developed further, and IDC was founded in 1967 as a separate organization centralizing portions of the input activities of its shareholders and providing them with searchable files (first on magnetic tape for in-house use, now online) [20.12]–[20.14] by extending its coverage to inorganics [20.15], polymers [20.16]–[20.18] (meanwhile discontinued), and by connecting it with the topological indexing and retrieval methods developed at BASF AG by ERNST MEYER et al. [19.52], [20.19].

Early in its development, GREMAS had already been designed to process patent literature. The detailed GREMAS fragmentation code, supplemented by syntactical input and retrieval tools, has proved to be a very effective search instrument for chemical patent literature [20.20], allowing searches with both high recall and high precision. GREMAS, being one of the first codes developed for reliable input and retrieval of generic structural formulae, has set a standard of performance which has not yet been matched by other more recent systems.

The inherent problems of the IDC system are the high indexing costs that have to be absorbed by the few share-holding companies and its steadily decreasing lead over the commercial services (Derwent, CAS).

20.5   INPADOC

The International Patent Documentation Center (INPADOC), Vienna, was founded in 1972 on the basis of an agreement between the WIPO and Austria [20.21]. It was integrated into the EPO on January 1, 1991. As the services of this organization have so far been connected with its former name, this is still used here; its new name is: European Patent Office, Principle Directorate Information. Strictly speaking, the INPADOC services should now be treated in Section 21.3.

The objective of INPADOC has been to collect the bibliographic data of patent documents of as many patent systems as possible (now more than 50 patent systems) and to process these into a range of services available to patent offices all over the world and to the public [20.22]. The services are issued as computer output on microfiche (COM), on magnetic tape, or are searchable on host computers at INPADOC, Maxwell Online, Dialog, STN International, and PATOLIS (Japan).

The files offered are:

– *Numerical Data bank*, containing the numbers of the documents processed in numerical order;
– *Patent Family Service*;
– *Patent Classification Service*, a register of IPC classes and the numbers of the patent documents to which these classes were allocated;
– *Patent Applicant Service*, a register of applicants' names in standardized form, and of the corresponding patent documents, also in priority order;
– *Patent Inventor Service*, a register of inventors' names and the corresponding patent documents;
– *Legal Status Service* [20.23];
– *International Patent Gazette*, a collection of the entries of the patent gazettes of all patent systems considered by INPADOC;
– additional specialized files.

Current awareness services (*Watch Services*) based on these files and searches on demand are also provided. These include searches performed in the PATOLIS data base of the Japanese Patent Office [20.24].

The INPADOC bibliographic data files have, particularly since the PCT and the European patent systems have come into full swing, become an indispensable, reliable resource for patent information management. The INPADOC files have been extensively used not only by patent offices and industrial enterprises, but also by commercial and private information services (e.g., CAS, Derwent, and IDC) by special arrangements.

20.6   Other Services

The other data bases and data banks listed in the WIPO Handbook (Section 20.1) are mostly confined to the documents of one or few national or supranational patent systems. The services of two further data base producers should be mentioned:

Claims Patent and Claims Citation. *IFI/Plenum Data Company* provides two large files: *Claims Patent* covers two million U.S. patent documents (chemistry since 1950, other areas since 1963) including design patents. Indexing is based on the most comprehensive claims of the documents, bibliographic data are included. The data base is searchable by general terms, fragment terms, CA Registry Numbers, IPC, and linear notation. In addition, IFI offers *Claims Citation*, containing the patent literature cited in U.S. patent documents since 1947, 5.5 million documents in all. Claims Patent is searchable at Pergamon Orbit Infoline, Dialog, and STN International, Claims Citation is searchable only on Dialog.

PATOS. *Bertelsmann Informations Service* and *Wila Verlag* jointly offer the PATOS online data base covering more than three million German, PCT, and European patent documents including German utility model (Gebrauchsmuster) documents. The data base is derived from the *Auszüge* (Section 18.3) as well as the European and PCT patent gazettes. PATOS is searchable on Bertelsmann's own host and can be retrieved in English and German and by IPC. The utility model documents and the PCT documents are covered since 1983, the other documents since 1986. PATOS also contains citations.

20.7 Shortcomings of Existing Services

A great many assessments and comparisons of existing patent information services were published over the years. Information managers are advised to treat the results presented and the theoretical conclusions drawn with caution. It is very difficult to compare complex information *services* in all their relevant aspects; however individual *searches* performed with the same query can be compared in two or three data bases. Whether the data presented reveal the "typical" performance of the data bases examined remains questionable; queries can be biassed, consciously or not [20.25]. Analyses are rarely performed over wide ranges of search strategies in big data bases because this is very expensive.

Nevertheless, practitioners know from day-to-day experience the shortcomings of the data bases they frequently use [19.2]–[19.6], [20.26]–[20.36]. The most important complaint is unsatisfactory data quality [19.7], [20.37]. *Lack of consistency* and *incompleteness of indexing*, even contrary to the services' own rules and standards, is a frequent cause of insufficient recall. Although it is difficult for the major data base producers to keep data quality high when thousands of documents in different layouts and languages must be processed weekly, more emphasis needs to be put on controlling and monitoring the input. The incompleteness of search results reported in the literature [19.5], [19.6], [19.8] is disquieting. Two data bases covering the same area of technology almost never give identical search results. It is disturbing to realize how little overlap is found.

Another pressing retrieval problem is *lack of precision*. The masses of irrelevant citations retrieved in normal searches in frequently used data bases is not only annoying but, in the long run, a tremendous waste of personell capacity and money on the user's side. Data base producers should make greater investments in improving their indexing languages and retrieval software.

The problem of *substructure searches* in generic formulae is discussed in Section 10.3. The users should not only encourage the data base producers to perfect their systems but also provide support and advice to them in this difficult matter.

20.8   "End-User" Searches

Structure searches of moderate complexity in patent literature, not meant to be complete, can be performed satisfactorily in CASONLINE by searchers not well versed with the system and using it infrequently (so-called end-users) [20.38] because it is a precision-type system based on topology. Although generalization is not possible, an inexperienced searcher should always seek the support of an expert in searches going beyond this limit, e.g., searches for nonstructural concepts or substructures.

Derwent's WPI, WPIL, and WPIM files clearly fall into the realm of information specialists. The same holds for searches for bibliographic patent data, e.g., in IN-PADOC files. Patent bibliography has become so complex that no "end-user" can come to terms with it.

It is helpful for an inexperienced searcher to discuss his information needs with a specialist and to formulate search strategies that he can modify and safely apply at his discretion.

21 Patent Information Services Provided by Governmental Authorities

21.1 Introduction

Growing dissatisfaction with the traditional role of the national patent offices, voiced not only by industry but also from within the patent offices, increasing "competition" by the supranational patent offices, and, last but not least, pressing internal needs for modernizing the offices' procedures have made governments more aware of the vital role patent information has played in national economies.

In recent years, services have been established that make the information tools of the patent offices available to the public.

The *European Patent Office* (EPO), having become prosperous by the enthusiastic, worldwide acceptance of its patent system, has made great investments in improving its internal information flow. The EPO has closely cooperated with the *United States Patent and Trademark Office* (USPTO) and with the *Japanese Patent Office* (JPO). Although these efforts are primarily directed to the patent offices' own purposes, the public has indirectly derived a benefit from these activities. The European Patent Office, in agreement with the governments being members of the European patent system, is prepared to pass the results of its developmental activities onto the public via the national patent offices. Moreover, the integration of INPADOC (Section 20.5) has provided the EPO with a direct outlet for its services.

21.2 Information from Patent Registers

Patent offices maintain extensive *patent registers* in which the data pertaining to every patent application received are meticulously noted. The patent offices offer services for answering information requests, the major offices allowing online access to their registers. These services are national alternatives to INPADOC's data banks with their worldwide coverage.

21.3   Retrieval Services

Numerous retrieval services that are accessible to the public have been established by the EPO and the national patent offices [21.1]–[21.6], only some are dealt with here.

The *EPAT* data bank giving full account of the European patents and published patent applications, including the Euro-PCT cases, has been available online at Télésystèmes-Questel since 1978 and covers more than 200 000 documents. This data bank is supplemented by *ECLATX*, also searchable online, a comprehensive explanation of the EPO's special version of the IPC (Section 19.3.3). EPAT and ECLATX are produced from EPO material by the French Patent Office (Institut National de la Propriété Industrielle, INPI). As a corollary of EPAT, INPI produces *FPAT*, a data base on French patents and patent applications.

The Directorate General 1 of the EPO, seated in The Hague, The Netherlands, already existed as Institut International des Brevets before the EPO was established and had offered patent searches in its large files to the public. This service continued after its integration into the EPO [21.7], [21.8]. *Standard searches*, identical in type and extent to the official searches provided for the European Patent Convention and the PCT, are offered as well as *special searches* tailored to meet the client's requirements.

The German Patent Office also carries out, on the basis of the German patent law, searches for users. Similar services are available from, e.g., the Austrian, British, Danish, French, Swedish, and Swiss patent offices. A particularly interesting data base offered by the Japanese Patent Office, PATOLIS, covering the Japanese patent and utility model documents can be accessed directly in Tokyo (knowledge of Japanese required) or via INPADOC [20.24].

PATDPA (Section 7.2.2), a data base established by the German Patent Office and Fachinformationszentrum Karlsruhe and searchable at STN International allows searches in German patent documents, including European and PCT documents designating Germany, and German utility model documents by IPC, descriptors, and free text. Abstracts (those printed on the title pages) and bibliographic data as well as graphical information transmitted via public telecommunication lines (narrow band-width) are displayable.

Many national patent offices have founded by themselves or in cooperation with other organizations decentralized patent document depositories in which manual searches can be performed in the stored documents or online searches can be ordered in public or commercial data bases.

22 Organizational Aspects of Patent Information Management in the Chemical Industry

22.1 The Position of Patent Information Management in the Enterprise

The organizational position of patent information management within an industrial enterprise is determined by the two-tier role of patent literature: the legal aspects call for allocating patent information management to the patent department, whereas the information aspects require its close coordination with research and development.

Patent information is much more frequently used by research personell than by patent specialists. It is for this reason and because up-to-date information retrieval facilities relying heavily on electronic data processing are more efficiently employed to capacity when centralized that preference should be given to integration of patent information management into overall corporate information management activities.

22.2 Cooperation Among Industrial Information Departments

22.2.1 Cooperation in Input

Years ago, all industrial companies had to manage the technical literature relevant to their activities by themselves. It was only logical to find ways and means for sharing the burden of indexing and cooperation schemes came into existence, especially in patent information management. The IDC (Section 20.4) also came into being by pooling certain input activities of its shareholders. Since indexing technical literature has become more and more the domain of commercial services, cooperation between industrial companies has been carried on in user groups.

22.2.2   User Groups and International Associations

User groups not only survey and criticize the commercial and public information providers' activities, they also test their new services and give competent advice on how to improve them. Examples of international patent information user groups are the *Patent Documentation Group* (PDG) [22.1], [22.2], which has existed since 1957, and the *International Association of Producers and Users of Online Patent Information* (OLPI), established in 1986 [22.3]. Cooperation of these organizations is being extended to patent offices. A considerable number of national (patent) information user groups also exist.

23   The Future of Patent Information Management

The future of patent information management is predominantly determined by developments taking place in three different fields: First in the area of patent laws, patent legislation and patent office procedures. Second in the area of science and industrial technology. Third in the area of computer technology, both hardware and software, and telecommunications.

The progress of science and technology constantly poses and solves new problems closely interlinked with many aspects of society. The proposed novel solutions are mirrored in the patent literature, and information management must respond to them in its indexing philosophy and its retrieval methods in order to offer its support effectively as a feedback process to industrial research and development.

A series of changes are ahead in the patent area as a result of new technologies (e.g., biotechnology), recent court decisions, new international agreements (e.g., on the EC patent), the increasing importance of the supranational patent systems, and political developments.

Computer and telecommunication technologies (e.g., online, CD-ROM) have revolutionized information management and can be expected to continue doing so [23.1]. Patent information managers must closely watch these changes, keep abreast with them, and react to them actively. This is a demanding task but also a great challenge for creative personalities.

This book can only present a snapshot of what information management is like today. It is the authors' pleasure to stimulate their readers to soon render it obsolete by making their own creative contributions to the fascinating world of information management.

24　References

References for Preface

[0.1] D. Bawden, K. Blakeman: *IT Strategies for Information Management*, Butterworths, London 1990.
[0.2] S. E. Arnold, *Online (Weston Conn.)* **15** (1991) no. 4, 39–51.
[0.3] H. Deyer, A. Gunson: *A Directory of Library and Information Retrieval Sofware for Micro-computers*, Gower, Aldershot 1990.
[0.4] E. N. Efthimiadis, *J. Doc.* **46** (1990) no. 3, 218–262.
[0.5] W. A. Warr, *Chemon. Intell. Lab. Syst.* **10** (1991) 279–292.
[0.6] D. Bawden, E. M. Mitchell (eds.): *Chemical Information Systems Beyond the Structure Diagram*, Ellis Horwood, Chichester, 1990.
[0.7] W. G. Town, *J. Chem. Inf. Comput. Sci.* **31** (1991) 176–180.
[0.8] W. A. Warr, *J. Chem. Inf. Comput. Sci.* **31** (1991) 181–186.
[0.9] J. E. Ash, W. A. Warr, P. Willet (eds.): *Chemical Structure Systems*, Ellis Horwood, Chichester, 1991.
[0.10] W. A. Warr, P. Willett, G. Downs (eds.): *Directory of Chemistry Software*, Cherwell Scientific Publishing, Oxford, 1991.

References for Chapter 1

[1.1] Y. Wolman: *Chemical Information. A Practical Guide to Utilization*, Wiley-Interscience, Chichester 1988.
[1.2] K. Subramanyam: *Scientific and Technical Information Resources*, Marcel Dekker, New York 1981.
[1.3] G. Wiggins: *Chemical Information Sources*, McGraw-Hill, New York 1990.
[1.4] R. E. Maizell: *How to Find Chemical Information A Guide for Practicing Chemists, Educators, and Students*, 2nd ed., John Wiley, New York 1987.
[1.5] A. Antony: *Guide to Basic Information Sources in Chemistry*, J. Wiley & Sons, Halsted Press, New York 1979.
[1.6] M. G. Mellon: *Chemical Publications, their Nature and Use*, 5th ed., McGraw-Hill, New York 1982.
[1.7] H. Skolnik: *The Literature Matrix of Chemistry*, Wiley-Interscience, New York 1982.
[1.8] M. Mücke: *Die Chemische Literatur. Ihre Erschließung und Benutzung*, VCH Verlagsgesellschaft, Weinheim 1982.
[1.9] A. J. Meadows (ed.): *The Scientific Journal*, Aslib, London 1979.
[1.10] A. J. Meadows (ed.): *The Growth of Science Publishing in Europe*, Elsevier, Amsterdam 1980.
[1.11] A. J. Meadows, *Libr. Review* **37** (1988) 7–16.
[1.12] H. Schulz: *From CA to CAS ONLINE*, VCH Verlagsgesellschaft, Weinheim 1988.
[1.13] J. M. Ziman: *Public Knowledge: The Social Dimension of Science*, Cambridge University Press, London 1968.
[1.14] J. R. Ravetz: *Scientific Knowledge and its Social Problems*, Penguin, Harmondsworth 1973;
[1.15] S. M. Dhawan, S. K. Phull, S. P. Jain, *J. Doc.* **36** (1980) no. 1, 24–41.
[1.16] E. Garfield: *Citation Indexing – Its Theory and Application in Science, Technology, and Humanities*, Wiley-Interscience, New York 1979.
[1.17] E. Garfield, *Science (Washington, D.C.)* **178** (1972) 471–479.
[1.18] R. Todorov, W. Glänzel, *J. Inf. Sci.* **14** (1988) 47–56.
[1.19] M. H. MacRoberts, B. R. MacRoberts, *J. Am. Soc. Inf. Sci.* **40** (1989) no. 5, 342–349.

[1.20] E. Garfield (ed.): *SCI Journal Citation Reports,* "A Bibliometric Analysis of Science Journals in the ISI Data Base," Institute for Scientific Information, Philadelphia 1987 (published annually).

[1.21] Chemical Abstracts Service Source Index (CASSI), *1907–1989 Cumulative Index,* Chemical Abstracts Service, Columbus, Ohio 1989.

[1.22] J. Hartley, M. Trueman, A. J. Meadows, *J. Inf. Sci.* **14** (1988) 69–75.

[1.23] R. F. Flesch, *J. Appl. Psychol.* **32** (1948) 221–233.

[1.24] G. R. Klare, *Reading Research Quarterly* **10** (1974–1975) 62–101.

[1.25] C. R. H. Inman, *J. Inf. Sci.* **6** (1983) 159–164.

[1.26] E. K. Samaha, *Information Development,* **3** (1987) no. 2, 103–107.

[1.27] H. H. Budzier, *Zentralbl. Bibliothekswesen* **100** (1986) no. 3, 93–101.

[1.28] W. Tuck, D. Archer, M.-C. Hayet, C. McKnight: *Project Quartet, LIR Report 76,* British Library, London 1990.

[1.29] W. Tuck, *Netlink* **5** (1989) no. 1, 5–8.

[1.30] R. M. Campbell, B. T. Stern, *Microcomputers for Information Management* **4** (1987) no. 2, 87–107.

[1.31] F. A. Mastroddi, J. Page in: *Electronic Publishing: State of the Art Report,* Pergamon Infotech, Maidenhead 1987, p. 37.

[1.32] B. F. Polansky, B. H. Weil, *J. Chem. Inf. Comput. Sci.* **25** (1985) 153–159.

[1.33] B. H. Weil, B. F. Polansky, *J. Chem. Inf. Comput. Sci.* **24** (1984) 43–50.

[1.34] B. F. Polansky in J. S. Dodd (ed.): *The ACS Style Guide,* American Chemical Society, Washington DC 1986, p. 137.

[1.35] D. P. Waite, *J. Chem. Inf. Comput. Sci.* **22** (1982) 63–66.

[1.36] P. J. Hills (ed.): *Trends in Information Transfer,* Francis Pinter, London 1982.

[1.37] L. C. Cross, *Aslib Proc.* **26** (1974) no. 11, 425–429.

[1.38] A. A. Manten, *J. Inf. Sci.* **1** (1980) 293–296.

[1.39] R. G. Lerner et al., *Annu. Rev. Inf. Sci. Technol.* **18** (1983) 127–149.

[1.40] D. P. Martinsen, R. A. Love, L. R. Garson, *Online (Weston Conn.)* **13** (1989) no. 2, 121–133.

[1.41] J. A. Hearty, *Information Services and Use* **8** (1988) 93–105.

References for Chapter 2

[2.1] M. Cooper, *J. Am. Soc. Inf. Sci.* **33** (1982) no. 3, 152–156.

[2.2] L. W. Granick, *J. Am. Soc. Inf. Sci.* **33** (1982) no. 3, 175–182.

[2.3] R. J. Rowlett, *J. Chem. Inf. Comput. Sci.* **25** (1985) no. 3, 159–163.

[2.4] B. H. Weil, *J. Am. Soc. Inf. Sci.* **21** (1970) 351–357.

[2.5] H. Schulz: *From CA to CAS ONLINE,* VCH Verlagsgesellschaft, Weinheim 1988.

[2.6] R. E. Maizell: *How to Find Chemical Information. A Guide for Practicing Chemists, Educators, and Students,* 2nd ed., John Wiley, New York 1987.

[2.7] D. B. Baker, J. W. Horiszny, W. V. Metanomski, *J. Chem. Inf. Comput. Sci.* **20** (1980) 193–201.

[2.8] H. L. Morgan, *J. Chem. Doc.* **5** (1965) 107–113.

[2.9] D. F. Zaye, W. V. Metanomski, A. J. Beach, *J. Chem. Inf. Comput. Sci.* **25** (1985) 392–399.

[2.10] M. Mücke: *Die Chemische Literatur. Ihre Erschließung und Benutzung,* VCH Verlagsgesellschaft, Weinheim 1982.

[2.11] R. S. Cahn, O. C. Dermer: *Introduction to Chemical Nomenclature,* 5th ed., Butterworths, London 1979.

[2.12] P. Fresenius: *Organic Chemical Nomenclature. Introduction to the Basic Principles,* Ellis Horwood, Chichester 1989.

[2.13] M. G. Robiette in R. Lees, A. Smith (ed.): *Chemical Nomenclature Usage,* Ellis Horwood, Chichester 1983, p. 74.

[2.14] *Nomenclature of Organic Chemistry, Sections A, B, C, D, E, F and H, (The Blue Book)*, Pergamon Press, Oxford 1979.

[2.15] *Nomenclature of Inorganic Chemistry, (The Red Book)*, 2nd ed., Butterworths, London 1971. [3rd ed. Blackwells Scientific, due January 1990].

[2.16] *Compendium of Analytical Nomenclature, (The Orange Book)*, 2nd ed., Blackwells Scientific, Oxford 1987.

[2.17] *Biochemical Nomenclature and Related Documents, (The Compendium)*, Biochemical Society, London 1978.

[2.18] N. Donaldson et al., *J. Chem. Doc.* **14** (1974) 3–14.

[2.19] G. G. Vander Stouw, I. Naznitsky, J. E. Rush, *J. Chem. Doc.* **7** (1967) 165–169.

[2.20] G. G. Vander Stouw, P. M. Elliott, A. C. Isenberg, *J. Chem. Doc.* **14** (1974) 185–193.

[2.21] G. G. Vander Stouw, *J. Chem. Inf. Comput. Sci.* **15** (1975) 232–236.

[2.22] D. I. Cooke-Fox, G. H. Kirby, J. D. Rayner, *J. Chem. Inf. Comput. Sci.* **29** (1989) 101–105.

[2.23] D. I. Cooke-Fox, G. H. Kirby, J. D. Rayner, *J. Chem. Inf. Comput. Sci.* **29** (1989) 106–112.

[2.24] D. I. Cooke-Fox, G. H. Kirby, J. D. Rayner, *J. Chem. Inf. Comput. Sci.* **29** (1989) 112–118.

[2.25] J. L. Wisniewski, *J. Chem. Inf. Comput. Sci.* **30** (1990) 324–332.

[2.26] P. G. Dittmar, R. E. Stobaugh, C. E. Watson, *J. Chem. Inf. Comput. Sci.* **20** (1980) 111–121.

[2.27] R. G. Freeland, S. A. Funk, L. J. O'Korn, G. A. Wilson, *J. Chem. Inf. Comput. Sci.* **19** (1979) 94–97.

[2.28] J. E. Blackwood, P. S. Elliott, R. E. Stobaugh, C. E. Watson, *J. Chem. Inf. Comput. Sci.* **17** (1977) 3–8.

[2.29] G. G. Vander Stouw, C. Gustafson, J. D. Rule, C. E. Watson, *J. Chem. Inf. Comput. Sci.* **16** (1976) 213–218.

[2.30] A. Zamora, D. L. Dayton, *J. Chem. Inf. Comput. Sci.* **16** (1976) 219–222.

[2.31] R. Stobaugh, *J. Chem. Inf. Comput. Sci.* **20** (1980) 76–82.

[2.32] J. Mockus, R. E. Stobaugh, *J. Chem. Inf. Comput. Sci.* **20** (1980) 18–22.

[2.33] J. P. Moosemiller, A. W. Ryan, R. E. Stobaugh, *J. Chem. Inf. Comput. Sci.* **20** (1980) 83–88.

[2.34] A. W. Ryan, R. E. Stobaugh, *J. Chem. Inf. Comput. Sci,* **22** (1982) 22–28.

[2.35] K. A. Hamill, R. D. Nelson, G. G. Vander Stouw, R. E. Stobaugh, *J. Chem. Inf. Comput. Sci.* **28** (1988) 175–179.

[2.36] R. E. Stobaugh, *J. Chem. Inf. Comput. Sci.* **28** (1988) 180–187.

[2.37] G. G. Vander Stouw in W. A. Warr (ed.): *Chemical Structures, The International Language of Chemistry*, Springer Verlag, Berlin 1988, p. 211.

[2.38] E. Garfield, *Science (Washington, D.C.)* **144** (1964) 649–654.

[2.39] E. Garfield, *J. Chem. Inf. Comput. Sci.* **25** (1985) 170–174.

[2.40] Y. Wolman: *Chemical Information. A Practical Guide to Utilization*, Wiley-Interscience, Chichester 1988.

[2.41] K. Subramanyam: *Scientific and Technical Information Resources*, Marcel Dekker, New York 1981.

[2.42] G. Wiggins: *Chemical Information Sources*. McGraw-Hill, New York 1990.

[2.43] A. Antony: *Guide to Basic Information Sources in Chemistry*, J. Wiley & Sons, Halsted Press, New York 1979.

[2.44] M. G. Mellon: *Chemical Publications, their Nature and Use*, 5th ed., McGraw-Hill, New York 1982.

[2.45] H. Skolnik: *The Literature Matrix of Chemistry*, Wiley-Interscience, New York 1982.

[2.46] J. Schmittroth (ed.): *Abstracting and Indexing Services Directory*, Gale Research, Detroit 1982–1983.

References for Chapter 3

[3.1] Y. Wolman: *Chemical Information. A Practical Guide to Utilization*, Wiley-Interscience, Chichester 1988.

[3.2] K. Subramanyam: *Scientific and Technical Information Resources*, Marcel Dekker, New York 1981.

[3.3] G. Wiggins: *Chemical Information Sources*, McGraw-Hill, New York 1990.

[3.4] R. E. Maizell: *How to Find Chemical Information. A Guide for Practicing Chemists, Educators, and Students*, 2nd ed., John Wiley, New York 1987.

[3.5] A. Antony: *Guide to Basic Information Sources in Chemistry*, J. Wiley & Sons, Halsted Press, New York 1979.

[3.6] M. G. Mellon: *Chemical Publications, their Nature and Use*, 5th ed., McGraw-Hill, New York 1982.

[3.7] H. Skolnik: *The Literature Matrix of Chemistry*, Wiley-Interscience, New York 1982.

[3.8] M. Mücke: *Die Chemische Literatur. Ihre Erschließung und Benutzung*, VCH Verlagsgesellschaft, Weinheim 1982.

[3.9] *Kirk–Othmer*, 3rd ed., 1978–1984.

[3.10] *Ullmann*, 4th ed., 1972–1984.

[3.11] *Ullmann*, 5th ed., 1985 to date.

[3.12] J. Matley, *Chem. Eng. Int. Ed.* **93** (1986) no. 8, 95–97.

[3.13] *Beilstein*, 1918 to date (now published in English).

[3.14] R. Luckenbach, R. Ecker, J. Sunkel, *Angew. Chem. Int. Ed. Engl.* **20** (1981) 841–849.

[3.15] R. Luckenbach, *J. Chem. Inf. Comput. Sci.* **21** (1982) 82–83.

[3.16] C. Jochum in W. A. Warr (ed.): *Chemical Structures, The International Language of Chemistry*, Springer Verlag, Berlin 1988, p. 187.

[3.17] *How to Use Beilstein*, Springer Verlag, Berlin 1984.

[3.18] A. J. Lawson in W. A. Warr (ed.): "Graphics for Chemical Structures: Integration with Text and Data," *ACS Symp. Ser.* **341** (1987) 80.

[3.19] S. R. Heller, (ed.): *The Beilstein Online Database. Implementation, Content and Retrieval*, ACS Symposion Series 436, American Chemical Society, Washington, D.C., 1990.

[3.20] H. O. House, *J. Chem. Inf. Comput. Sci.* **24** (1984) 277.

[3.21] *Gmelin*, 1922 to date.

[3.22] J. Buckingham (ed.): *Dictionary of Organic Compounds*, 5th ed., Chapman and Hall, London 1982 and annual supplements thereafter.

[3.23] J. Buckingham, *CHEMTECH* **15** (1985) no. 11, 674–679.

[3.24] P. Hyams, *Inf. World Rev.* **1989**, July, 18.

[3.25] *Houben-Weyl*, 4th ed., 1952 to date.

[3.26] W. Theilheimer: *Synthetic Methods of Organic Chemistry*, Karger, Basel 1946 to date.

[3.27] A. F. Finch, *J. Chem. Inf. Comput. Sci.* **26** (1986) no. 1, 17–22.

[3.28] A. F. Finch in P. Willett (ed.): *Modern Approaches to Chemical Reaction Searching*, Gower, Aldershot 1986, p. 36.

[3.29] *Organic Syntheses*, 2nd ed., Wiley, New York 1941 to date.

[3.30] W. Dauben (ed.): *Organic Reactions*, Wiley, New York, 1942 to date.

[3.31] L. G. Wade (ed.): *Compendium of Organic Synthetic Methods*, Wiley, New York 1977 to date (Volumes 1 and 2 were edited by I. T. Harrison and S. Harrison).

[3.32] M. Fieser (ed.): *Fieser and Fieser's Reagents for Organic Synthesis*, Wiley, New York 1967 to date.

[3.33] *Inorganic Syntheses*, Wiley, New York 1939 to date.

[3.34] S. Coffey or M. F. Ansell (ed.): *Rodd's Chemistry of Carbon Compounds*, 2nd ed., Elsevier, Amsterdam 1964 to date.

[3.35] D. Barton, W. D. Ollis (eds.): *Comprehensive Organic Chemistry*, Pergamon Press, Oxford 1979.

[3.36] G. Wilkinson (ed.): *Comprehensive Organometallic Chemistry*, Pergamon Press, Oxford 1982.

[3.37] *The Chemistry of Heterocyclic Compounds – A Series of Monographs*, Wiley, New York 1970 to date.

[3.38] A. R. Katritzky, C. W. Rees (eds.): *Comprehensive Heterocyclic Chemistry*, Pergamon Press, Oxford 1984.

[3.39] G. Wilkinson (ed.): *Comprehensive Coordination Chemistry*, Pergamon Press, Oxford 1987.

[3.40] J. C. Bailar, H. J. Emeleus, R. Nyholm, A. F. Trotman, (eds.): *Comprehensive Inorganic Chemistry*, Pergamon Press, Oxford 1973.

[3.41] S. Budavari (ed.): *The Merck Index*, 11th ed., Merck and Company, Rahway, New Jersey 1989.

[3.42] R. C. Weast, J. G. Grasselli (eds.): *CRC Handbook of Data on Organic Compounds*, 2nd ed., CRC Press, Boca Raton, Florida 1988.

[3.43] N. I. Sax, R. J. Lewis (eds.): *Dangerous Properties of Industrial Materials*, 7th ed., Van Nostrand Reinhold, New York 1989.

[3.44] L. Bretherick: *Handbook of Reactive Chemical Hazards*, 4th ed., Butterworths, London 1985.

[3.45] L. Bretherick (ed.): *Hazards in the Chemical Laboratory*, Royal Society of Chemistry, London 1986.

References for Chapter 4

[4.1] D. N. Wood, *IFLA J.* **10** (1984) no. 3, 278–282.

[4.2] N. W. Posnett, W. J. Baulkwill, *J. Inf. Sci.* **5** (1982) 121–130.

[4.3] V. Alberani, *Orv. Konyvtaros* **28** (1988) no. 4, 341–350.

[4.4] N. W. Posnett, *Libr. Acquisitions: Practice Theory* **8** (1984) 275–285.

[4.5] J. P. Chillag in D. F. Shaw (ed.): *Information Sources in Physics*, Butterworths, London 1985, p. 355.

[4.6] C. Hasemann, *Z. Bibliothekswesen Bibliographie* **33** (1986) no. 6, 417–427.

[4.7] D. N. Wood, *Aslib Proc.* **34** (1982) no. 11/12, 459–465.

[4.8] J. P. Chillag in D. W. Bromley, A. M. Allott (eds.): *British Librarianship and Information Work 1981–1985*, vol. 2: "Special Libraries, Materials and Processes," Library Association, London 1988, p. 95.

[4.9] V. Alberani, A. Pagamonci (eds.): "Letteratura Grigia," *Boll. Inf. Assoc. Ital. Biblioteche* **27** (1987) no. 3/4, 305–498.

[4.10] J. M. Gibb, M. Maurice, *Aslib Proc.* **34** (1982) no. 11/12, 493–497.

[4.11] C. Hasemann, *ABI-Tech.* **5** (1985) no. 4, 261–265.

[4.12] C. Salmon, L. van Simaeys, *Cah. Doc.* **3** (1980) 53–56.

[4.13] E. K. Samaha, *Inf. Dev.* **3** (1987) no. 2, 103–107.

[4.14] J. S. Robinson: *Tapping the Government Grapevine: The User-Friendly Guide to US Government Information Services*, Oryx Press, Phoenix 1988.

[4.15] H. Ogawa et al., *J. Am. Soc. Inf. Sci.* **40** (1989) no. 5, 350–355.

[4.16] E. Kohl, M. Ockenfeld: *Konferenzinformation: Hinweise zur Ermittlung und Beschaffung von Terminen und Vorträgen chemierelevanter Veranstaltungen*, Arbeitsgruppe Informationswissenschaft in der Chemie an der Universität Frankfurt am Main (AIC), Frankfurt 1981.

[4.17] *World Meetings: United States and Canada*, MacMillan, New York 1963, quarterly.

[4.18] *World Meetings: Outside United States and Canada*, MacMillan, New York 1968, quarterly.

[4.19] *World Meetings: Medicine*, MacMillan, New York 1978, quarterly.

[4.20] M. Ockenfeld: *Dissertationen als Informationsquellen. Hinweise zu ihrer Ermittlung, Beschaffung, Gestaltung und Verbreitung im Fach Chemie*, Arbeitsgruppe Informationswissenschaft in der Chemie an der Universität Frankfurt am Main (AIC), Frankfurt 1981.

References for Chapter 5

[5.1] *Croner's A–Z of Business Information*, Croner, Kingston-upon-Thames, UK 1989.
[5.2] S. Ball: *Directory of International Sources of Business Information*, Pitman, London 1989.
[5.3] S. P. Webb: *Where to Buy Business Information*, Headland Press, Cleveland 1990.
[5.4] P. and A. Foster (eds.): *The Online Business Sourcebook*, Headland Press, Headland, UK 1989.
[5.5] *Directory of Online Databases*, Cuandra Associates, Santa Monica, CA, issued quarterly.
[5.6] P. Millard: *Trade Associations and Professional Bodies of the UK*, Pergamon Press, Oxford 1969.
[5.7] D. Lasok, J. W. Bridge: *Introduction to the Law and Institutions of the European Communities*, Butterworth, Seven Oaks, UK 1982.
[5.8] J. Jeffries: *Guide to the Official Publications of the European Communities*, Mansell, London 1981.

References for Chapter 6

[6.1] P. F. Burton, J. H. Petrie: *The Librarian's Guide to Microcomputers for Information Management*, Van Nostrand Reinhold (UK), Wokingham 1986.
[6.2] P. Zorkoczy: *Information Technology An Introduction*, Pitman, London 1985.
[6.3] P. C. Treleven, I. G. Lima, *Computer* **15** (1982) no. 8, 79–88.
[6.4] Her Majesty's Stationery Office (HMSO): *A Programme for Advanced Information Technology: the Report of the Alvey Committee*, HMSO, London 1982.
[6.5] P. Bishop: *Fifth Generation Computers*, Ellis Horwood, Chichester 1986.
[6.6] P. Salenieks: *Computing: The Next Generation*, Ellis Horwood, Chichester 1988.
[6.7] A. Peled, *Sci. Am.* **257** (1987) no. 4, 35–42.
[6.8] J. D. Meindl, *Sci. Am.* **257** (1987) no. 4, 54–62.
[6.9] G. C. Fox, P. C. Messina, *Sci. Am.* **257** (1987) no. 4, 44–52.
[6.10] J. D. Foley, *Sci. Am.* **257** (1987) no. 4, 82–90.
[6.11] D. Bawden, *J. Inf. Sci.* **11** (1985) 1–8.
[6.12] R. C. Rouse, *Program* **23** (1989) no. 3, 269–275.
[6.13] W. T. Wipke in W. A. Warr (ed.): "Graphics for Chemical Structures Integration with Text and Data," *ACS Symp. Ser.* **341** (1987).
[6.14] A. E. Cawkell, *The Electronic Library* **7** (1989) no. 1, 24–28.
[6.15] A. E. Cawkell, *The Electronic Library* **7** (1989) no. 2, 106–110.
[6.16] A. E. Cawkell, *The Electronic Library* **7** (1989) no. 3, 180–184.
[6.17] *Monitor* **42** (1984) 8.
[6.18] M. Rivett, *J. Inf. Sci.* **13** (1987) 25–34.
[6.19] D. H. Davies, *J. Am. Soc. Inf. Sci.* **39** (1988) no. 1, 34–42.
[6.20] E. M. Cichocki, S. M. Ziemer, *J. Am. Soc. Inf. Sci.* **39** (1988) no. 1, 43–46.
[6.21] T. Hendley, *Inf. Media Technology* **19** (1986) no. 3, 103–106.
[6.22] M. S. White, *World Patent Information* **8** (1986) no. 3, 177–181.
[6.23] L. B. Glass, *Byte* **14** (1989) no. 5, 283–289.
[6.24] S. Lambert, S. Ropiequet: *CD-ROM: The New Papyrus*, Microsoft Press, Redmond, WA 1986.
[6.25] C. Oppenheim (ed.): *CD-ROM Fundamentals to Applications*, Butterworth, London 1988.
[6.26] C. Sherman: *The CD-ROM Handbook*, McGraw-Hill, New York 1988.
[6.27] M. H. Kryder, *Sci. Am.* **257** (1987) no. 4, 72–81.
[6.28] *Electronic and Optical Publishing Review* **8** (1988) no. 2, 102–103.
[6.29] D. Pountain, *Byte* **14** (1989) no. 2, 274–280.
[6.30] G. A. Pierce, *Proc. Int. Soc. Optical Engineering* **899** (1988) 31–33.

[6.31] R. B. Barnes, F. J. Sukernick, *J. Information and Image Management* **19** (1986) no. 10, 34–38.
[6.32] G. I. Ouchi: *Personal Computers for Scientists*, American Chemical Society, Washington, DC 1986.
[6.33] R. Alberico: *Microcomputers for the Online Searcher*, Meckler, Westport, CT 1987.

References for Chapter 7

[7.1] P. J. Denning, R. L. Brown, *Sci. Am.* **251** (1984) 80–89.
[7.2] N. Wirth, *Sci. Am.* **251** (1984) 48–57.
[7.3] L. G. Tesler, *Sci. Am.* **251** (1984) 58–66.
[7.4] M. Lesk, *Annu. Rev. Inf. Sci. Technol.* **19** (1984) 97–128.
[7.5] T. Oren, G. A. Kildall, *IEEE Spectrum* **4** (1986) 49–54.
[7.6] L. A. Kurtz, *Program* **18** (1984) 1–15.
[7.7] F. D. Gault in J. R. Humble, V. E. Hampel (eds.): *Data Base Management in Science and Technology*, Elsevier, Amsterdam 1984.
[7.8] J. H. Ashford, *Program* **18** (1984) 16–45.
[7.9] R. Kimberley (ed.): *Text Retrieval A Directory of Software*, 3rd ed., Gower, Aldershot 1990.
[7.10] P. F. Burton, H. Gates, *Program* **19** (1985) 1–19.
[7.11] J. H. Ashford, *Program* **18** (1984) 124–146.
[7.12] W. A. Warr in D. E. Meyer, W. A. Warr, R. A. Love (eds.): *Chemical Structure Software for Personal Computers*, American Chemical Society, Washington, DC, 1988, p.37.
[7.13] B. W. Boehm, *Computer* **20** (1987) no. 9, 43–57.
[7.14] I. Somerville: *Software Engineering*, Addison-Wesley, Wokingham 1985.
[7.15] D. M. Jennings et al., *Science (Washington, D.C.)* **231** (1986) 943–950.

References for Chapter 8

[8.1] R. J. Hartley, E. M. Keen, J. A. Large, L. A. Tedd: *Online Searching: Principles and Practice*, Bowker-Saur, London 1990.
[8.2] J. Convey, C. Bingley: *Online Information Retrieval. An Introductory Manual to Principles and Practice*, 3rd ed., Library Association, London 1989.
[8.3] G. Turpie: *Going Online, 1988*, Aslib, London 1988.
[8.4] C. H. Fenichel, T. H. Hogan: *Online Searching: A Primer*, Learned Information, Marlton, NJ, 1989.
[8.5] H. Stack: *Online Searching Made Simple*, PJB Publications, London 1988.
[8.6] *Online Searching in Science and Technology: An Introductory Guide to Equipment and Search Techniques*, British Library Science Reference and Information Service, London 1988.
[8.7] R. K. Summit in A. Kent, H. Lancourt (eds.): *Encyclopedia of Library and Information Science*, vol. 7, Marcel Dekker, New York 1972.
[8.8] C. A. Cuadra, *J. Chem. Inf. Comput. Sci.* **15** (1975) 48–51.
[8.9] W. A. Warr, A. R. Haygarth Jackson, *J. Chem. Inf. Comput. Sci.* **28** (1988) 68–72.
[8.10] D. T. Hawkins, L. R. Levy, *Online (Weston, Conn.)* **9** (1985) no. 6, 30–36.
[8.11] M. Morrison, *Online (Weston, Conn.)* **13** (1989) no. 4, 46–52.
[8.12] R. V. Janke, *Online (Weston, Conn.)* **7** (1983) no. 5, 12–29.
[8.13] K. Y. Marcaccio (ed.): *Computer-Readable Data bases: A Directory and Data Sourcebook*, Gale Research, Detroit 1989.
[8.14] *Directory of Online Databases*, Cuadra Associates, Santa Monica, CA, issued quarterly.
[8.15] D. M. Cipra, C. F. Damron, *Database* **8** (1985) no. 2, 23–30.
[8.16] R. L. M. Synge, *J. Chem. Inf. Comput. Sci.* **30** (1990) no. 1, 33–35.
[8.17] D. T. Hawkins, *Database* **8** (1985) no. 2, 31–41.

[8.18] Y. Wolman: *Chemical Information. A Practical Guide to Utilization*, Wiley-Interscience, Chichester 1988.

[8.19] R. E. Maizell: *How to Find Chemical Information A Guide for Practicing Chemists, Educators, and Students*, 2nd ed., John Wiley, New York 1987.

[8.20] M. Mücke: *Die Chemische Literatur. Ihre Erschließung und Benutzung*, VCH Verlagsgesellschaft, Weinheim 1982.

[8.21] A. B. Piternick, *Online Rev.* **13** (1989) no. 6, 457–476.

[8.22] H. Schulz: *From CA to CAS ONLINE*, VCH Verlagsgesellschaft, Weinheim 1988.

[8.23] J. Witiak, *Database* **11** (1988) no. 2, 95–96.

[8.24] K. D. Lehmann, H. Strohl-Goebel (eds.): *The Application of Microcomputers in Information, Documentation and Libraries*, Elsevier North-Holland, Amsterdam 1987.

[8.25] P. Leggate, H. Dyer, *The Electronic Library* **4** (1986) no. 1, 38–49.

[8.26] S. J. Kolner, *Online (Weston, Conn.)* **9** (1985) no. 1, 37–42.

[8.27] S. J. Kolner, *Online (Weston, Conn.)* **9** (1985) no. 2, 39–46.

[8.28] S. J. Kolner, *Online (Weston, Conn.)* **9** (1985) no. 3, 44–50.

[8.29] S. J. Kolner, *Online (Weston, Conn.)* **9** (1985) no. 4, 27–34.

[8.30] S. J. Kolner, *Online (Weston, Conn.)* **9** (1985) no. 6, 42–50.

[8.31] S. J. Kolner, *Online (Weston, Conn.)* **10** (1986) no. 4, 32–36.

[8.32] P. F. Burton, J. H. Petrie: *The Librarian's Guide to Microcomputers for Information Management*, Van Nostrand Reinhold (UK), Wokingham 1986.

[8.33] P. Nieuwenhuysen, *Online Rev.* **11** (1987) no. 6, 363–367.

[8.34] P. Nieuwenhuysen, *The Electronic Library* **6** (1988) no. 3, 168–172.

[8.35] L. R. Levy, D. T. Hawkins, *Online (Weston, Conn.)* **10** (1986) no. 1, 33–40.

[8.36] D. T. Hawkins, L. R. Levy, *Online (Weston, Conn.)* **10** (1986) no. 3, 49–58.

[8.37] W. A. Warr, *Database* **10** (1987) no. 3, 122–128.

[8.38] R. Walsh, *Aslib Information* **16** (1988) nos. 11/12, 282–283.

[8.39] S. Cisler, *Online (Weston, Conn.)* **12** (1988) no. 6, 99–102.

[8.40] *Monitor* **91** (1988) 5–8.

[8.41] C. Oppenheim, *Advanced Information Report*, March 1990, 7–9.

[8.42] C. A. Kehoe, *Online Rev.* **9**(1985) no. 6, 489–505.

[8.43] *The Electronic Library* **4** (1986) no. 1, 30–33.

[8.44] T. A. Hanson, *Aslib Proc.* **41** (1989) no. 9, 267–274.

[8.45] N. Hoyle, *Database* **10** (1987) no. 1, 73–78.

[8.46] A. N. Grosch, *Online Rev.* **12** (1988) no. 6, 375–386.

[8.47] C. Rodwell, P. Clayton, *Advanced Information Report* **10** (1988) no. 8, 9–11.

[8.48] J. K. Pemberton, *Online (Weston, Conn.)* **10** (1986) no. 3, 17–24.

[8.49] C. Tenopir, *Library J.* **1986**, 48–49.

[8.50] C. Tenopir, *Annu. Rev. Inf. Sci. Technol.* **19** (1984) 215–246.

[8.51] R. Summit, A. Lee, *Serials Rev.* **3** (1988) 7–10.

[8.52] J. A. Hearty, *Information Services and Use* **8** (1988) 93–105.

[8.53] *Kirk–Othmer*, 3rd ed., 1978–1984.

[8.54] E. W. Johnson, M. P. Kutz, *Electronic Publishing and Bookselling* **2** (1984) no. 1, 17–19.

[8.55] P. Willett, *Information Processing and Management* **24** (1988) 577–597.

[8.56] P. Willett (ed.): *Document Retrieval Systems*, Taylor Graham, London 1988.

[8.57] R. G. Lerner et al., *Annu. Rev. Inf. Sci. Technol.* **18** (1983) 127–149.

[8.58] D. P. Martinsen, R. A. Love, L. R. Garson, *Online (Weston Conn.)* **13** (1989) no. 2, 121–133.

[8.59] J. A. Hearty, *Information Services and Use* **8** (1988) 93–105.

[8.60] G. Tittlbach: "State of the Art Report," in: *Electronic Publishing*, Pergamon, Oxford 1987, p. 91.

[8.61] G. Tittlbach, *Nachr. Dok.* **37** (1986) nos. 4–5, 198–204.

[8.62] G. Tittlbach, *J. Chem. Inf. Comput. Sci.* **26** (1986) no. 1, 13–17.

[8.63] W. Detemple, *Online Rev.* **13** (1989) no. 2, 155–160.

[8.64] W. Niedermeyr in W. A. Warr (ed.): *Graphics for Chemical Structures. Integration with Text and Data*, ACS Symposium Series 341, American Chemical Society, Washington, D.C., 1987, p. 143.

[8.65] N. J. Thompson, *Online (Weston, Conn.)* **13** (1989) no. 3, 15–26.

[8.66] T. B. Chadwick, *Online (Weston, Conn.)* **13** (1989) no. 3, 28–30.

[8.67] J. Buckingham (ed.): *Dictionary of Organic Compounds*, 5th ed., Chapman and Hall, London 1982 and annual supplements thereafter.

[8.68] P. Hyams, *Information World Review*, July 1989, 18.

[8.69] W. Tuck, D. Archer, M.-C. Hayet, C. McKnight: *Project Quartet, LIR Report 76*, British Library, London 1990.

[8.70] W. Tuck, *Netlink* **5** (1989) no. 1, 5–8.

[8.71] R. M. Campbell, B. T. Stern, *Microcomputers for Information Management* **4** (1987) no. 2, 87–107.

[8.72] F. A. Mastroddi, J. Page in: *Electronic Publishing: State of the Art Report*, Pergamon Infotech, Maidenhead 1987, p. 37.

[8.73] A. Buscain, *Aslib Proc.* **37** (1985) nos. 6/7, 249–256.

[8.74] D. D. Baird, *Aslib Proc.* **37** (1985) nos. 6/7, 257–265.

[8.75] A. J. Metcalf, *Aslib Proc.* **37** (1985) nos. 6/7, 267–271.

[8.76] C. H. Jacobs, *Aslib Proc.* **37** (1985) nos. 6/7, 273–276.

[8.77] P. N. Hunter, *Aslib Proc.* **37** (1985) nos. 6/7, 277–280.

[8.78] W. R. Tuck, *Aslib Proc.* **38** (1985) no. 3, 85–92.

[8.79] R. Veith, *Annu. Rev. Inf. Sci. Technol.* **18** (1983) 3–28.

[8.80] R. Kimberley (ed.): *Text Retrieval A Directory of Software*, 3rd ed., Gower, Aldershot 1990.

[8.81] M. M. K. Hlava (ed.): Bulletin of the American Society for Information Science, Oct/Nov 1987, 14–27.

[8.82] C. Oppenheim (ed.): *CD-ROM Fundamentals to Applications*, Butterworth, London 1988.

[8.83] C. Sherman: *The CD-ROM Handbook*, McGraw-Hill, New York 1988.

[8.84] S. E. Arnold, L. Rosen: *Managing the New Electronic Information Products*, Riverside Data, Sudbury, MA, 1989.

[8.85] J. P. Roth (ed.): *CD-ROM Applications and Markets*, Meckler, Westport, CT 1988.

[8.86] J. Mitchell, J. Harrison (eds.): *The CD-ROM Directory 1990*, TFPL Publishing, London 1989.

[8.87] N. Desmaris (ed.): *CD-ROMs in Print 1990: An International Guide*, Meckler, Westport, CT, 1991.

[8.88] S. Oberlin, J. Cox: *The Microsoft CD-ROM Yearbook (1989–1990)*, Microsoft Press, Redmond, WA 1989.

[8.89] J. M. Burridge, *Biochem. Soc. Trans.* **17** (1989) 840–841.

[8.90] I. A. Penn et al.: *Records Management Handbook*, AIIM Publications, Silver Spring, MD 1989.

[8.91] P. Emmerson (ed.): *How to Manage Your Records: A Guide to Effective Practice*, ICSA, Cambridge 1989.

[8.92] S. James: *Records Management: An Introduction*, TFPL Publishing, London 1989.

[8.93] 1989 International Micrographics Source Book Including Related Imaging Technologies, Microfilm Publishing, Larchmont, NY 1989.

[8.94] R. J. Focarelli et al.: *The Microform Connection*, AIIM Publications, Silver Spring, MD, 1989.

[8.95] A. Shiel: *Optical Disk Storage and Document Image Processing: A Guide and Directory*, 2nd ed., Cimtech, Hatfield 1990.

[8.96] J. P. Roth, B. A. Berg: *Software for Optical Storage*, AIIM Publications, Silver Spring, MD, 1989.

[8.97] W. Saffady: *Optical Storage Technology 1989: A State of the Art Review*, Meckler, Westport, CT, 1989.

[8.98] A. E. Cawkell, *The Electronic Library* **7** (1989) no. 1, 24–28.

[8.99] A. E. Cawkell, *The Electronic Library* **7** (1989) no. 2, 106–110.

[8.100] A. E. Cawkell, *The Electronic Library* **7** (1989) no. 3, 180–184.

[8.101] A. E. Cawkell, *The Electronic Library* **7** (1989) no. 4, 248–250.

[8.102] A. E. Cawkell, *The Electronic Library* **7** (1989) no. 5, 317–323.

[8.103] R. G. Lerner et al., *Annu. Rev. Inf. Sci. Technol.* **18** (1983) 127–149.

[8.104] W. G. Town, *J. Chem. Inf. Comput. Sci.* **31** (1991) 176–180.

[8.105] W. Saffady: *Optical Disks versus Micrographics as Document Storage and Retrieval Technologies*, Meckler, Westport, Connecticut 1988.

[8.106] G. Salton, M. J. McGill: *Introduction to Modern Information Retrieval*, McGraw-Hill Computer Science Series, New York 1983.

[8.107] K. Sparck-Jones, *Information Processing and Management* **24** (1988) no. 6, 703–711.

[8.108] K. Sparck-Jones, J. I. Tait, *J. Doc.* **40** (1984) no. 1, 50–66.

[8.109] I. G. Hendry, P. Willett, F. E. Wood, *Program* **20** (1986) 245–263.

[8.110] I. G. Hendry, P. Willett, F. E. Wood, *Program* **20** (1986) 382–393.

[8.111] S. J. Wade, P. Willett, *Program* **22** (1988) 44–61.

[8.112] G. Salton, *Commun. ACM* **29** (1986) 648–656.

[8.113] G. Salton: *Automatic Text Processing: The Transformation, Analysis and Retrieval of Information*, Addison-Wesley, Reading 1989.

[8.114] G. Salton, C. Buckley, *J. Am Soc. Inf. Sci.* **41**, no. 4 (1990) 288–297.

[8.115] K. Sparck-Jones, *J. Inf. Sci.* **1** (1980) 325–332.

[8.116] C. W. Cleverdon, *Information Services and Uses* **4** (1984) 37–47.

[8.117] P. Willett, *J. Inf. Sci.* **15** (1989) nos. 4/5, 223–236.

References for Chapter 9

[9.1] J. E. Ash et al.: *Communication, Storage and Retrieval of Chemical Information*, Ellis Horwood, Chichester 1985.

[9.2] J. M. Barnard in C. Citroen, J. M. Griffith (eds.): *Perspectives in Information Management*, vol. 1, Butterworths, Guildford 1989.

[9.3] R. E. Maizell: *How to Find Chemical Information*, 2nd ed., Wiley, New York 1987, pp. 152–200.

[9.4] W. A. Warr (ed.): *Graphics for Chemical Structures: Integration with Text and Data*, ACS Symposium Series 341, American Chemical Society, Washington, DC, 1987.

[9.5] W. A. Warr (ed.): *Chemical Structures, The International Language of Chemistry*, Springer Verlag, Berlin 1988.

[9.6] D. E. Meyer, W. A. Warr, R. A. Love (eds.): *Chemical Structure Software for Personal Computers*, American Chemical Society, Washington, DC, 1988.

[9.7] W. A. Warr in W. A. Warr (ed.): *Chemical Structure Information Systems: Interfaces, Communication and Standards*, ACS Symposium Series 400, American Chemical Society, Washington, DC, 1989, p. 1.

[9.8] J. Figueras in B. W. Rossiter (ed.): *Physical Methods of Chemistry*, vol. 1: "Components of Scientific Instruments and Applications of Computers to Chemical Research," Wiley, New York 1986, p. 687.

[9.9] J. Buckingham (ed.): *Dictionary of Organic Compounds*, 5th ed., Chapman and Hall, London 1982 and annual supplements thereafter.

[9.10] H. P. Kollig, *Toxicol. Environ. Chem.* **17** (1988) 287–311.

[9.11] R. Luckenbach, R. Ecker, J. Sunkel, *Angew. Chem. Int. Ed. Engl.* **20** (1981) 841–849.

[9.12] S. R. Heller (ed.): *The Beilstein Ouline Database: Implementation, Content, and Retrieval*, ACS Symposion Series 436, American Chemical Society, Washington, D.C., 1990.

[9.13] D. R. Lide, *Science (Washington, D. C)* **212** (1981) 1343–1349.

[9.14] F. H. Allen et al., *J. Appl. Crystallogr.* **7** (1974) 73–78.

[9.15] S. R. Heller, K. Scott, D. W. Bigwood, *J. Chem. Inf. Comput. Sci.* **29** (1989) 159–162.

[9.16] R. E. Maizell: *How to Find Chemical Information*, 2nd ed., Wiley, New York 1987, pp. 327–330.

[9.17] C. A. Shelley in J. Zupan (ed.): *Computer-Supported Spectroscopic Databases*, Ellis Horwood, Chichester 1986, p. 6.

[9.18] G. W. A. Milne et al., *Org. Mass Spectrom.* **17** (1982) no. 11, 547–552.

[9.19] J. R. Rumble, D. R. Lide, *J. Chem. Inf. Comput. Sci.* **25** (1985) 231–235.

[9.20] S. R. Heller, *J. Chem. Inf. Comput. Sci.* **25** (1985) 224–231.

[9.21] S. R. Heller, G. W. A. Milne, R. J. Feldmann, *Science (Washington, D. C)* **195** (1977) 253–259.

[9.22] G. W. A. Milne, C. L. Fisk, S. R. Heller, R. Potenzone, *Science (Washington, D. C)* **215** (1982) 371–375.

[9.23] O. Yamamoto et al., *Anal. Sci.* **4** (1988) no. 3, 233–239.

[9.24] S. R. Lowry, D. A. Huppler, C. R. Anderson, *J. Chem. Inf. Comput. Sci.* **25** (1985) 235–241.

[9.25] P. R. Griffiths, C. L. Wilkins, *Appl. Spectrosc.* **42** (1988) no. 4, 538–545.

[9.26] S. R. Heller, *Anal. Chem.* **44** (1972) 1951–1961.

[9.27] S. R. Heller, R. J. Feldmann, H. M. Fales, G. W. A. Milne, *J. Chem. Doc.* **13** (1973) 130–133.

[9.28] R. S. Heller, G. W. A. Milne, R. J. Feldmann, S. R. Heller, *J. Chem. Inf. Comput. Sci.* **16** (1976) 176–178.

[9.29] S. R. Heller, R. S. Heller, D. P. Martinsen, *Adv. Mass Spectrom.* **8B** (1980) 1578–1581.

[9.30] S. R. Heller in J. Zupan (ed.): *Computer-Supported Spectroscopic Databases*, Ellis Horwood, Chichester 1986, p. 1 and p. 118.

[9.31] D. D. Speck, R. Venkataraghavan, F. W. McLafferty, *Org. Mass Spectrom.* **13** (1978) no. 4, 209–213.

[9.32] F. W. McLafferty, R. H. Hertel, R. D. Villwock, *Org. Mass Spectrom.* **9** (1974) no. 7, 690–702.

[9.33] G. M. Pesyna, R. Venkataraghavan, H. E. Dayringer, F. W. McLafferty, *Anal. Chem.* **48** (1976) 1362–1368.

[9.34] I. K. Mun, R. Venkataraghavan, F. W. McLafferty, *Anal. Chem.* **49** (1977) 1723–1726.

[9.35] B. L. Atwater (Fell), R. Venkataraghavan, F. W. McLafferty, *Anal. Chem.* **51** (1979) 1945–1949.

[9.36] F. W. McLafferty et al., *Int. J. Mass Spectrom. Ion Phys.* **47** (1983) 317–319.

[9.37] S. Anderson, *J. Mol. Graphics* **2** (1984) 83–90.

[9.38] N. A. Farmer, M. P. O'Hara, *Database* **3** (1980) 10–25.

[9.39] D. E. Meyer in W. A. Warr (ed.): *Chemical Structures: The International Language of Chemistry*, Springer Verlag, Berlin 1988, p. 251.

[9.40] D. del Rey in W. A. Warr (ed.): *Graphics for Chemical Structures: Integration with Text and Data*, ACS Symposium Series 341, American Chemical Society, Washington, D.C. 1987, p. 48.

[9.41] W. G. Town, *Chem. Br.* **25** (1989) no. 11, 1118–1120.

[9.42] H. S. Hertz, R. A. Hites, K. Biemann, *Anal. Chem.* **43** (1971) 681–691.

[9.43] B. L. Atwater, D. B. Stauffer, F. W. McLafferty, D. W. Peterson, *Anal. Chem.* **57** (1985) 899–903.

[9.44] A. Hanna, J. C. Marshall, T. L. Isenhour, *J. Chromatogr. Sci.* **17** (1979) 434–440.

[9.45] M. D. Erickson, *Appl. Spectrosc.* **35** (1981) 181–184.

[9.46] C. W. Small, G. T. Rasmussen, T. L. Isenhour, *Appl. Spectrosc.* **33** (1979) 444–450.

[9.47] M. F. Delaney, P. C. Uden, *Anal. Chem.* **51** (1979) 1242–1243.

[9.48] J. A. de Haseth, L. V. Azarraga, *Anal. Chem.* **53** (1981) 2292–2296.

[9.49] G. W. Milne, S. R. Heller, *J. Chem. Inf. Comput. Sci.* **20** (1980) 204–211.

[9.50] S. R. Lowry, D. A. Huppler, *Anal. Chem.* **53** (1981) 889–893.

[9.51] S. R. Lowry, D. A. Huppler, *Anal. Chem.* **55** (1983) 1288–1291.

[9.52] R. Schwarzenbach, J. Meili, H. Koenitzer, J. T. Clerc, *Org. Magn. Reson.* **8** (1976) no. 1, 11–16.

[9.53] A. B. Wagner, *Online Rev.* **10** (1986) no. 3, 173–183.

[9.54] M. Passlack, W. Bremser in J. Zupan (ed.): *Computer-Supported Spectroscopic Databases*, Ellis Horwood, Chichester 1986, p. 92.

[9.55] W. Bremser, *Angew. Chem. Int. Ed. Engl.* **27** (1988) 247–260.

[9.56] W. Bremser, R. Neudert, *Eur. Spectrosc. News* **75** (1987) 10–27.

[9.57] R. Neudert, W. Bremser, H. Wagner, *Org. Mass Spectrom.* **22** (1987) 321–329.

[9.58] H. Damen, D. Henneberg, B. Weimann, *Anal. Chim. Acta* **103** (1978) 289–302.

[9.59] H. Kalchhauser, W. Robien, *J. Chem. Inf. Comput. Sci.* **25** (1985) 103–108.

[9.60] *Crystallographic Databases*, International Union of Crystallography, Chester 1987.

[9.61] O. Kennard et al., *Chem. Br.* **11** (1975) 213–216.

[9.62] C. Kratky, W. Robien, *Österr. Chem. Z.* **89** (1988) no. 3, 58–62.

[9.63] S. Bellard in: *Crystallographic Databases*, International Union of Crystallography, Chester 1987, p. 39.

[9.64] F. H. Allen, J. E. Davies, J. J. Galloy, O. Johnson et al., *J. Chem. Inf. Comput. Sci.* **31** (1991) 187–204.

[9.65] E. E. Abola, F. C. Bernstein, T. F. Koetzle, in P. J. Glaeser (ed.): *The Role of Data in Scientific Progress*, Elsevier, New York 1985.

[9.66] I. D. Brown, *Acta Crystallogr. Sect. A* **A41** (1985) 399.

[9.67] A. D. Mighell, V. L. Himes, *Acta Crystallogr. Sect. A* **A42** (1986) 101–105.

[9.68] G. Bergerhoff, R. Hundt, R. Sievers, I. D. Brown, *J. Chem. Inf. Comput. Sci.* **23** (1983) 66–69.

[9.69] M. J. E. Sternberg, S. A. Islam, *Biochem. Soc. Trans.* **17** (1989) 845–847.

[9.70] H. W. Mewes, A. Elzanowski, D. G. George, *Biochem. Soc. Trans.* **17** (1989) 843–845.

[9.71] *Blaise Newsletter* No. 93 (1988) 19–20.

[9.72] CODATA Conference on Scientific and Technical Data in a New Era, 26–29 Sept. 1988, Karlsruhe, FRG.

[9.73] R. E. Buntrock, *Database* **13** (1990) no. 3, 99–100.

[9.74] S. R. Heller (ed.): *The Beilstein Online Database: Implementation, Content, and Retrieval*, ACS Symposium Series 436, American Chemical Society, Washington 1990.

[9.75] D. J. Huddart, *Aslib Proc.* **40** (1988) no. 5, 133–137.

[9.76] D. T. Hawkins, *Database* **8** (1985) no. 2, 31–41.

[9.77] F. C. Allan, W. R. Ferrell, *Database* **12** (1989) no. 3, 50–58.

[9.78] C. Dutheuil, *Newsidic* **91** (1988) 9–12.

[9.79] *Online Rev.* **14** (1990) no. 1, 54.

[9.80] *The Electronic Library* **6** (1988) no. 5, 381.

[9.81] *DECHEMA Chemistry Data Series*, Deutsche Gesellschaft für Chemisches Apparatewesen, Chemische Technik und Biotechnologie e.V. , Frankfurt, 10 volumes.

[9.82] D. Bawden, *Aslib Proc.* **40** (1988) no. 3, 79–85.

[9.83] F. E. Wood, A. T. Berrie, H. R. Plampin, M. L. Wilkinson-Tough, *J. Inf. Sci.* **15** (1989) 269–276.

[9.84] O. Norager in W. A. Warr (ed.): *Chemical Structures: The International Language of Chemistry*, Springer Verlag, Berlin 1988.

[9.85] S. R. Heller, *J. Chem. Inf. Comput. Sci.* **29** (1989) no. 2, 135–136.

[9.86] Y. Wolman: *Chemical Information. A Practical Guide to Utilization*, Wiley, Chichester 1983, Chap. 6.

[9.87] S. V. Meschel, *Online Rev.* **8** (1984) no. 1, 77–101.

References for Chapter 10

[10.1] W. A. Warr in R. Lees, A. Smith (eds.): *Chemical Nomenclature Usage*, Ellis Horwood, Chichester 1983, p.124.

[10.2] P. Willett, *J. Chemom.* **1** (1987) 139–155.

[10.3] J. Ash et al., (eds.): *Communication, Storage and Retrieval of Chemical Information*, Ellis Horwood, Chichester 1985, Chap. 5, p. 128.

[10.4] J. Figueras in B. W. Rossiter (ed.): *Physical Methods of Chemistry*, vol. 1: "Components of Scientific Instruments and Applications of Computers to Chemical Research," Wiley, New York 1986, p. 687.

[10.5] D. Bawden, T. Devon, *Database* **3** (1980) no. 3, 29–39.

[10.6] S. M. Kaback, *J. Chem. Inf. Comput. Sci.* **20** (1980) 1–6.

[10.7] R. Fugmann in J. E. Ash, E. Hyde (eds.): *Chemical Information Systems*, Ellis Horwood, Chichester 1975, p. 195.

[10.8] C. Suhr, E. von Harsdorf, W. Dethlefsen in J. M. Barnard (ed.): *Computer Handling of Generic Chemical Structures*, Gower Pub. Co., Aldershot–Brookfield 1984, p.10.

[10.9] U. Schoch-Grübler, *Online Rev.* **14** (1990) no. 2, 95–108.

[10.10] P. A. Baker, G. Palmer, P. W. L. Nichols in J. E. Ash, E. Hyde (eds.): *Chemical Information Systems*, Ellis Horwood, Chichester 1975, p.97.

[10.11] W. J. Wiswesser, *J. Chem. Inf. Comput. Sci.* **22** (1982) 88–93.

[10.12] G. Palmer, *Chem. Br.* **6** (1970) 422.

[10.13] G. M. Dyson in J. E. Ash, E. Hyde (eds.): *Chemical Information Systems*, Ellis Horwood, Chichester 1975, p.130.

[10.14] C. E. Granito, M. D. Rosenberg, *J. Chem. Doc.* **11** (1971) 251–256. E. Garfield, M. Sim, *Pure Appl. Chem.* **49** (1977) 1803.

[10.15] D. R. Eakin in J. E. Ash, E. Hyde (eds.): *Chemical Information Systems*, Ellis Horwood, Chichester 1975, p.227.

[10.16] D. R. Eakin, E. Hyde, G. Palmer, *Pestic. Sci.* **5** (1974) 319–326.

[10.17] E. E. Townsley, W. A. Warr in W. J. Howe, M. M. Milne, A. F. Pennell (eds.): *Retrieval of Medicinal Chemical Information*, ACS Symposium Series 84, American Chemical Society, Washington, D.C., 1978, p.73.

[10.18] W. A. Warr in: *Proceedings of the 5th International Online Information Meeting*, Learned Information, Oxford 1981, p.391.

[10.19] W. A. Warr, *J. Mol. Graphics* **4** (1986) 165–169.

[10.20] D. J. Weininger, *J. Chem. Inf. Comput. Sci.* **28** (1988) 31–36.

[10.21] J. E. Ash in J. E. Ash, E. Hyde (eds.): *Chemical Information Systems*, Ellis Horwood, Chichester 1975, p.156.

[10.22] H. Bebak et al., *J. Chem. Inf. Comput. Sci.* **29** (1989) 1–5.

[10.23] J. M. Barnard, *J. Chem. Inf. Comput. Sci.* **30** (1990) 81–96.

[10.24] G. Moreau, *Nouv. J. Chim.* **4** (1980) 17–22.

[10.25] W. T. Wipke, T. M. Dyott, *J. Am. Chem. Soc.* **96** (1974) 4825–4834.

[10.26] M. F. Lynch, J. Orton, W. G. Town, *J. Chem. Soc. C* **1969**, 1732–1736.

[10.27] J. H. R. Bragg, M. F. Lynch, W. G. Town, *J. Chem. Doc.* **10** (1970) 125–128.

[10.28] R. G. Freeland, S. A. Funk, L. J. O'Korn, G. A. Wilson, *J. Chem. Inf. Comput. Sci.* **19** (1979) 94–97.

[10.29] W. T. Wipke, S. K. Krishnan, G. I. Ouchi, *J. Chem. Inf. Comput. Sci.* **18** (1978) 32–37.

[10.30] D. Bawden et al., *J. Chem. Inf. Comput. Sci.* **21** (1981) 83–86.

[10.31] J. M. Barnard in W. A. Warr (ed.): *Chemical Structures, The International Language of Chemistry*, Springer Verlag, Berlin 1988, p.113.

[10.32] M. F. Lynch in J. E. Ash, E. Hyde (eds.): *Chemical Information Systems*, Ellis Horwood, Chichester 1975, p.177.

[10.33] J. E. Crowe, M. F. Lynch, W. G. Town, *J. Chem. Soc. C* **1970**, 990–996.

[10.34] G. W. Adamson, M. F. Lynch, W. G. Town, *J. Chem. Soc. C* **1971**, 3702–3706.

[10.35] G. W. Adamson et al., *J. Chem. Doc.* **13** (1973) 153–157.

[10.36] G. W. Adamson, S. E. Creasey, J. P. Eakins, M. F. Lynch, *J. Chem. Soc. Perkin Trans.* **1973**, 2071–2076.

[10.37] W. Graf, H. K. Kaindl, H. Kniess, R. Warszawski, *J. Chem. Inf. Comput. Sci.* **22** (1982) 177–181.

[10.38] P. G. Dittmar et al., *J. Chem. Inf. Comput. Sci.* **23** (1983) 93–102.

[10.39] R. Attias, *J. Chem. Inf. Comput. Sci.* **23** (1983) 102–108.

[10.40] L. C. Ray, R. A. Kirsch, *Science (Washington, D. C.)* **126** (1957) 814–819.

[10.41] S. H. Unger, *Commun. ACM* **7** (1964) 26–34.

[10.42] E. H. Sussenguth, *J. Chem. Doc.* **5** (1965) 36–43.

[10.43] J. Figueras, *J. Chem. Doc.* **12** (1972) 237–244.

[10.44] A. von Scholley, *J. Chem. Inf. Comput. Sci.* **24** (1984) 235–241.

[10.45] V. J. Gillet et al., *J. Chem. Inf. Comput. Sci.* **26** (1986) 118–126.

[10.46] S. R. Heller, *J. Chem. Inf. Comput. Sci.* **25** (1985) 224–231.

[10.47] G. W. A. Milne, C. L. Fisk, S. R. Heller, R. Potenzone, *Science (Washington D.C.)* **215** (1982) 371–375.

[10.48] G. W. Milne, S. R. Heller, *J. Chem. Inf. Comput. Sci.* **20** (1980) 204–211.

[10.49] S. R. Heller, G. W. A. Milne, R. J. Feldmann, *Science (Washington, D. C.)* **195** (1977) 253–259.

[10.50] R. J. Feldmann, *J. Chem. Inf. Comput. Sci.* **17** (1977) 157–163.

[10.51] G. W. A. Milne et al., *J. Chem. Inf. Comput. Sci.* **18** (1978) no. 4, 181–185.

[10.52] M. Z. Nagy, S. Kozics, T. Veszpremi, P. Bruck in W. A. Warr (ed.): *Chemical Structures, The International Language of Chemistry*, Springer Verlag, Berlin 1988, p.127.

[10.53] J. M. Barnard (ed.): *Computer Handling of Generic Chemical Structures*, Gower Pub. Co., Aldershot–Brookfield 1984.

[10.54] J. M. Barnard, *Database* **10** (1987) no. 3, 27–34.

[10.55] M. F. Lynch, *World Patent Inf.* **8** (1986) 85–91.

[10.56] M. F. Lynch, J. M. Barnard, S. M. Welford, *J. Chem. Inf. Comput. Sci.* **25** (1985) 264–270.

[10.57] G. Downs, V. Gillet, J. Holliday, M. F. Lynch in W. A. Warr (ed.): *Chemical Structures, The International Language of Chemistry*, Springer Verlag, Berlin 1988, p.151.

[10.58] J. M. Barnard, *J. Chem. Inf. Comput. Sci.* **21** (1981) 151–161.

[10.59] J. M. Barnard, M. F. Lynch, S. M. Welford, *J. Chem. Inf. Comput. Sci.* **24** (1984) 66–71.

[10.60] J. M. Barnard, M. F. Lynch, S. M. Welford, *J. Chem. Inf. Comput. Sci.* **22** (1982) 160–164.

[10.61] S. M. Welford, M. F. Lynch, J. M. Barnard, *J. Chem. Inf. Comput. Sci.* **21** (1981) 161–168.

[10.62] S. M. Welford, M. F. Lynch, J. M. Barnard, *J. Chem. Inf. Comput. Sci.* **24** (1984) 66–71.

[10.63] V. J. Gillet et al., *J. Chem. Inf. Comput. Sci.* **27** (1987) 126–137.

[10.64] American Chemical Society, US 4 642 762, 1987 (W. Fisanick).

[10.65] T. Nakayama, Y. Fujiwara, *J. Chem. Inf. Comput. Sci.* **23** (1983) 80–87.

[10.66] C. Fricke, I. Nickelsen, R. Fugmann, J. Sander, *Tetrahedron Comput. Methodol.* **2** (1989) no. 3, 167–175.

[10.67] K. E. Shenton, P. Norton, E. A. Fearns in W. A. Warr (ed.): *Chemical Structures, The International Language of Chemistry*, Springer Verlag, Berlin 1988, p.169.

[10.68] W. Fisanick, *J. Chem. Inf. Comput. Sci.* **30** (1990) 145–154.

[10.69] F. H. Allen, M. F. Lynch, *Chem. Br.* **25** (1989) no. 11, 1101–1104, 1108.

[10.70] P. Gund, *Prog. Mol. Subcell. Biol.* **5** (1977) 117–143.

[10.71] Y. C. Martin, M. G. Bures, P. Willett in K. B. Lipkowitz, D. B. Boyd (eds.): *Reviews in Computational Chemistry*, VCH, New York 1990, p. 213–256.

[10.72] R. P. Sheridan et al., *J. Chem. Inf. Comput. Sci.* **29** (1989) 255–260.

[10.73] A. T. Brint, E. Mitchell, P. Willett in W. A. Warr (ed.): *Chemical Structures, The Internation-al Language of Chemistry*, Springer Verlag, Berlin 1988, p.131.

[10.74] S. E. Jakes, P. Willett, *J. Mol. Graphics* **4** (1986) 12–20.

[10.75] A. T. Brint, P. Willett, *J. Mol. Graphics* **5** (1987) 49–56.

[10.76] A. T. Brint, P. Willett, *J. Chem. Inf. Comput. Sci.* **27** (1987) 152–158.

[10.77] C. W. Crandall, D. H. Smith, *J. Chem. Inf. Comput. Sci.* **23** (1983) 186–197.

[10.78] A. Rusinko III et al., *J. Chem. Inf. Comput. Sci.* **29** (1989) 251–255.

[10.79] J. H. Van Drie, D. Weininger, Y.C. Martin, *J. Comput.-Aided Mol. Des.* **3** (1989) no. 3, 225–251.

[10.80] R. E. Carhart, D. H. Smith, R. Venkataraghavan, *J. Chem. Inf. Comput. Sci.* **25** (1985) 64–73.

[10.81] P. Willett, V. Winterman, D. Bawden, *J. Chem. Inf. Comput. Sci.* **26** (1986) 36–41.

[10.82] D. Bawden in W. A. Warr (ed.): *Chemical Structures, The International Language of Chem-istry*, Springer Verlag, Berlin 1988, p. 145.

[10.83] P. Willett: *Similarity and Clustering in Chemical Information Systems*, Research Studies Press, Letchworth 1987.

[10.84] P. Willett, V. Winterman, D. Bawden, *J. Chem. Inf. Comput. Sci.* **26** (1986) 109–118.

[10.85] N. Farmer et al. in W. A. Warr (ed.): *Chemical Structures, The International Language of Chemistry*, Springer Verlag, Berlin 1988, p. 283.

[10.86] C. Jochum, T. Worbs in W. A. Warr (ed.): *Chemical Structures, The International Language of Chemistry*, Springer Verlag, Berlin 1988, p. 279.

[10.87] P. Willett, *J. Inf. Sci.* **15** (1989) nos. 4/5, 223–236.

[10.88] C. A. Pogue, E. M. Rasmussen, P. Willett, *Parallel Computing* **8** (1988) 399–407.

[10.89] E. M. Rasmussen, G. M. Downs, P. Willett, *J. Comput. Chem.* **9** (1988) no. 4, 378–386.

[10.90] H. M. Grindley et al. in H. R. Collier (ed.): *Chemical Information, Information in Chem-istry, Pharmacology and Patents*, Springer Verlag, Berlin 1989, p. 253.

[10.91] P. J. Artymiuk, D. W. Rice, E. M. Mitchell, P. Willett, *J. Inf. Sci.* **15** (1989) 287–298.

[10.92] W. T. Wipke, D. Rogers, *J. Chem. Inf. Comput. Sci.* **24** (1984) 255–262.

[10.93] A. T. Brint, P. Willett, *J. Mol. Graphics* **5** (1987) 200–207.

[10.94] A. T. Brint et al., *Parallel Computing* **8** (1988) 295–300.

[10.95] W. T. Wipke in W. A. Warr (ed.): "Graphics for Chemical Structures Integration with Text and Data," *ACS Symp. Ser.* **341** (1987).

[10.96] W. J. Howe, T. R. Hagadone, *J. Chem. Inf. Comput. Sci.* **22** (1982) 8–15.

[10.97] W. J. Howe, T. R. Hagadone, *J. Chem. Inf. Comput. Sci.* **22** (1982) 182–186.

[10.98] D. Bawden, *Chem. Br.* **25** (1989) no. 11, 1107–1108.

[10.99] T. R. Hagadone in W. A. Warr (ed.): *Chemical Structures, The International Language of Chemistry*, Springer Verlag, Berlin 1988, p. 23.

[10.100] W. G. Town, *Chem. Br.* **25** (1989) no. 11, 1118–1120.

[10.101] W. G. Town in W. A. Warr (ed.): *Chemical Structures, The International Language of Chemistry*, Springer Verlag, Berlin 1988, p. 243.

[10.102] W. A. Warr (ed.): *Chemical Structure Information Systems: Interfaces, Communication and Standards*, ACS Symposium Series 400, American Chemical Society, Washington, DC, 1989.

[10.103] N. A. Farmer, M. P. O'Hara, *Database* **3** (1980) 10–25.

[10.104] Y. Wolman: *Chemical Information. A Practical Guide to Utilization,* Wiley-Interscience, Chichester 1988.

[10.105] H. R. Pichler in H. Schulz (ed.): *From CA to CAS Online*, VCH, Weinheim 1988, p. 146.

[10.106] S. V. Kasparek: *Computer Graphics and Chemical Structures*, Wiley-Interscience, New York 1990.

[10.107] R. Buntrock, *Database* **11** (1988) no. 1, 87–88.

[10.108] J. Mockus, R. E. Stobaugh, *J. Chem. Inf. Comput. Sci.* **20** (1980) 18–22.

[10.109] H. R. Pichler, *Chem. Labor Betr.* **34** (1983) no. 5, 188–196.

[10.110] U. Jordis, O. Oberhauser, *Österr. Chem. Z.* **1982**, 311–314.

[10.111] S. R. Heller in: *Proceedings of the 7th International Online Information Meeting*, Learned Information, Oxford 1983, p. 81.

[10.112] A. Meurling, *Database* **13** (1990) no. 1, 54–63.

[10.113] M. G. Hicks, C. Jochum, *J. Chem. Inf. Comput. Sci.* **30** (1990) 191–199.

[10.114] J. Ash et al., (eds.): *Communication, Storage and Retrieval of Chemical Information*, Ellis Horwood, Chichester 1985, Chap. 7, p. 182.

[10.115] A. P. Lurie in W. A. Warr (ed.): *Chemical Structures, The International Language of Chemistry*, Springer Verlag, Berlin 1988, p. 77.

[10.116] R. E. Maizell: *How to Find Chemical Information A Guide for Practicing Chemists, Educators, and Students,* 2nd ed., John Wiley, New York 1987.

[10.117] A. B. Wagner, *Online Rev.* **10** (1986) no. 3, 173–183.

[10.118] G. W. A. Milne in W. A. Warr (ed.): *Graphics for Chemical Structures: Integration with Text and Data*, ACS Symposium Series 341, American Chemical Society, Washington, DC, 1987, p. 102.

[10.119] J. R. McDaniel, A. E. Fein in W. A. Warr (ed.): *Graphics for Chemical Structures: Integration with Text and Data*, ACS Symposium Series 341, American Chemical Society, Washington, DC, 1987, p. 62.

[10.120] F. H. Allen, J. E. Davies, J. J. Galloy, O. Johnson et al., *J. Chem. Inf. Comput. Sci.* **31** (1991) 187–204.

[10.121] J. E. Dubois, H. Viellard, *Bull. Soc. Chim. Fr.* **1968**, 900–919.

[10.122] J. E. Dubois, J. P. Anselmini, M. Chastrette, F. Hennequin, *Bull. Soc. Chim. Fr.* **1969**, 2439–2448.

[10.123] J. E. Dubois, D. Laurent, *Bull. Soc. Chim. Fr.* **1969**, 2449–2455.

[10.124] J. E. Dubois, H. Viellard, *Bull. Soc. Chim. Fr.* **1971**, 839–848.

[10.125] J. E. Dubois, *J. Chem. Doc.* **13** (1973) 8–13.

[10.126] G. Bauer in: *Proceedings of the 5th International Online Information Meeting*, Learned Information, Oxford 1981, p. 377.

[10.127] J.-P. Gay, G. Auneveux, F. Chabernaud in W. A. Warr (ed.): *Chemical Structure Information Systems, Interfaces, Communication and Standards*, ACS Symposium Series 400, American Chemical Society, Washington, DC, 1989, p. 89.

[10.128] J.-P. Gay, H. Alardo in H. R. Collier (ed.): *Chemical Information: Information in Chemistry, Pharmacology and Patents*, Springer Verlag, Berlin 1989, p. 221.

[10.129] A. J. C. M. de Jong, A. M. C. Deibel in W. A. Warr (ed.): *Chemical Structures, The International Language of Chemistry*, Springer Verlag 1988, p. 45.

[10.130] P. Bruck, M. Z. Nagy, S. Kozics in: *Proceedings of the 11th International Online Information Meeting*, Learned Information, Oxford 1987, p. 41.

[10.131] S. R. Heller in: *Proceedings of the 11th International Online Information Meeting*, Learned Information, Oxford 1987, p. 25.

[10.132] S. Anderson, *J. Mol. Graphics* **2** (1984) 83–90.

[10.133] G. W. Adamson, J. M. Bird, G. Palmer, W. A. Warr, *J. Chem. Inf. Comput. Sci.* **25** (1985) 90–92.

[10.134] G. W. Adamson, J. M. Bird, G. Palmer, W. A. Warr, *J. Mol. Graphics* **4** (1986) 165–169.

[10.135] S. Barcza, L. A. Kelly, S. S. Wahrman, R. E. Kirschenbaum, *J. Chem. Inf. Comput. Sci.* **25** (1985) 55–59.

[10.136] S. Barcza, H. W. Mah, M. H. Myers, S. S. Wahrman, *J. Chem. Inf. Comput. Sci.* **26** (1986) 198–204.

[10.137] T. M. Johns in W. A. Warr (ed.): *Graphics for Chemical Structures: Integration with Text and Data*, ACS Symposium Series 341, American Chemical Society, Washington, DC 1987, p. 18.

[10.138] A. P. Lurie in W. A. Warr (ed.): *Chemical Structures, The International Language of Chemistry*, Springer Verlag, Berlin 1988, p. 77.

[10.139] W. A. Warr in: *Proceedings of the 7th International Online Information Meeting*, Learned Information, Oxford 1983, p. 91.

[10.140] T. Legatt, A. Saltzman, *Drug Inf. J.* **20** (1986) 51–56.

[10.141] W. T. Wipke, T. M. Dyott, *J. Am. Chem. Soc.* **96** (1974) 4834–4842.

[10.142] A. J. Stuper, P. C. Jurs, *J. Chem. Inf. Comput. Sci.* **16** (1976) 99–105.

[10.143] A. J. Stuper, W. E. Brugger, P. C. Jurs: *Computer-Assisted Studies of Chemical Structure and Biological Function*, Wiley, New York 1979.

[10.144] D. Magrill in W. A. Warr (ed.): *Chemical Structures, The International Language of Chemistry*, Springer Verlag, Berlin 1988, p. 53.

[10.145] L. Domokos, C. Jochum, H. Maier in H. R. Collier (ed.): *Chemical Information: Information in Chemistry, Pharmacology and Patents*, Springer Verlag, Berlin 1989, p. 191.

[10.146] D. Bawden et al. in W. A. Warr (ed.): *Chemical Structures, The International Language of Chemistry*, Springer Verlag, Berlin 1988, p. 63.

[10.147] J. Kao, V. Day, L. Watt, *J. Chem. Inf. Comput. Sci.* **25** (1985) no. 2, 129–135.

[10.148] H. K. Kaindl in H. R. Collier (ed.): *Chemical Information: Information in Chemistry, Pharmacology and Patents*, Springer Verlag, Berlin 1989, p. 63.

[10.149] D. E. Meyer, W. A. Warr, R. A. Love (eds.): *Chemical Structure Software for Personal Computers*, American Chemical Society, Washington, DC, 1988.

[10.150] J. F. Barstow, D. del Rey, J. S. Laufer, *Am. Lab. (Fairfield, Conn.)* **20** (1988) no. 7, 82–85.

[10.151] C. K. Gerson, R. A. Love, *Anal. Chem.* **59** (1987) no. 17, 1031A-1048A.

[10.152] D. E. Meyer in W. A. Warr (ed.): *Chemical Structures: The International Language of Chemistry*, Springer Verlag, Berlin 1988, p. 251.

[10.153] D. del Rey in W. A. Warr (ed.): *Graphics for Chemical Structures: Integration with Text and Data*, ACS Symposium Series 341, American Chemical Society, Washington, DC, 1987, p. 48.

[10.154] C. Seiter, P. Cohan, *Int. Lab.* **17** (1987) no. 7, 62–67.

[10.155] W. G. Town, *Chem. Br.* **25** (1989) no. 11, 1118–1120.

[10.156] W. G. Town in: *Proceedings of the 11th International Online Information Meeting*, Learned Information, Oxford 1987, p. 33.

[10.157] W. A. Warr, M. P. Wilkins, *Online (Weston, Conn.)* **14** (1990) no. 3, 50–54.

[10.158] J. M. Barnard, C. J. Jochum, S. M. Welford in W. A. Warr (ed.): *Chemical Structure Information Systems, Interfaces, Communication, and Standards*, ACS Symposium Series 400, American Chemical Society, Washington, DC, 1989, p. 76.

[10.159] S. R. Heller, *Database* **10** (1987) no. 4, 47–52.

[10.160] R. Franke: *Theoretical Drug Design Methods*, Elsevier, Amsterdam 1984.

[10.161] J. G. Topliss (ed.): *Quantitative Structure-Activity Relationships of Drugs*, Academic Press, New York 1983.

[10.162] Y. C. Martin, *J. Med. Chem.* **24** (1981) 229–237.

[10.163] J. L. Fauchere (ed.): *QSAR: Quantitative Structure-Activity Relationships in Drug Design*, Alan R. Liss, New York 1989.

[10.164] S. Borman, *Chem. Eng. News*, Feb. 19, 1990, 20–23.

[10.165] S. M. Free, J. W. Wilson, *J. Med. Chem.* **7** (1964) 395–399.

[10.166] T. Fujita, T. Ban, *J. Med. Chem.* **14** (1971) 148–152.

[10.167] C. Hansch, *Acc. Chem. Res.* **2** (1969) 232–239.

[10.168] S. H. Unger: *Consequences of the Hansch Paradigm for the Pharmaceutical Industry*, Academic Press, New York 1980.

[10.169] C. Hansch, *Drug Dev. Res.* **1** (1981) 267–309.

[10.170] C. Hansch, A. J. Leo: *Substituent Constants for Correlation Analysis in Chemistry and Biology*, Wiley Interscience, New York 1979.

[10.171] Y. C. Martin: *Quantitative Drug Design. A Critical Introduction*, Marcel Dekker, New York 1978.

[10.172] Y. C. Martin in E. J. Ariens (ed.): *Drug Design*, Academic Press, New York 1979, p. 8.

[10.173] P. C. Jurs, T. L. Isenhour: *Chemical Applications of Pattern Recognition*, Wiley, New York 1975.

[10.174] P. C. Jurs, T. R. Stouch, M. Czerwinski, J. N. Narvaez, *J. Chem. Inf. Comput. Sci.* **25** (1985) 296–308.

[10.175] D. Bawden, *J. Chem. Inf. Comput. Sci.* **23** (1983) 14–22.

[10.176] R. D. Cramer, G. Redl, C. E. Berkoff, *J. Med. Chem.* **17** (1974) 533–535.

[10.177] L. Hodes, G. F. Hazard, R. I. Geran, S. Richman, *J. Med. Chem.* **20** (1977) 469–475.

[10.178] L. Hodes, *J. Chem. Inf. Comput. Sci.* **21** (1981) 128–136.

[10.179] E. Meyer, E. Sens, in W. A. Warr (ed.): *Chemical Structures: The International Language of Chemistry,* Springer, Berlin 1988.

[10.180] G. W. Adamson, D. Bawden, *J. Chem. Inf. Comput. Sci.* **15** (1975) 215–220.

[10.181] G. W. Adamson, D. Bawden, *J. Chem. Inf. Comput. Sci.* **16** (1976) 161–165.

[10.182] G. W. Adamson, D. Bawden, *J. Chem. Inf. Comput. Sci.* **17** (1977) 164–171.

[10.183] G. W. Adamson, D. Bawden, *J. Chem. Inf. Comput. Sci.* **20** (1980) 97–100.

[10.184] G. W. Adamson, D. Bawden, D. T. Saggers, *Pestic. Sci.* **15** (1984) 31–39.

[10.185] G. W. Adamson, J. A. Bush, *Nature (London)* **248** (1974) 406–407.

[10.186] G. W. Adamson, J. A. Bush, *J. Chem. Soc. Perkin Trans.* **1** (1976) 168–172.

[10.187] R. Langridge, T. E. Ferrin, I. D. Kuntz, M. L. Connolly, *Science (Washington, D. C.)* **211** (1981) 661–666.

[10.188] C. Humblet, G. R. Marshall, *Drug Dev. Res.* **1** (1981) 409–434.

[10.189] P. Gund, J. D. Andose, J. B. Rhodes, G. M. Smith, *Science (Washington, D. C.)* **208** (1980) 1425–1431.

[10.190] A. J. Hopfinger, *J. Med. Chem.* **28** (1985) 1133–1139.

[10.191] N. C. Cohen, *Adv. Drug Res.* **14** (1985) 42–145.

[10.192] S. H. Unger, *Drug Inf. J.* **21** (1987) no. 3, 267–275.

[10.193] G. W. A. Milne, J. S. Driscoll, V. E. Marquez in H. R. Collier (ed.): *Chemical Information: Information in Chemistry, Pharmacology and Patents*, Springer Verlag, Berlin 1989, p. 19.

[10.194] J. G. Vinter, M. Harris, *Chem. Br.* **25** (1989) no. 11, 1111–1116.

[10.195] P. Willett: *Modern Approaches to Chemical Reaction Searching*, Gower, Aldershot 1988.

[10.196] G. E. Vladutz, *Inf. Storage Retr.* **1** (1963) 117–146.

[10.197] P. Willett, *J. Chem. Inf. Comput. Sci.* **20** (1980) 93–96.

[10.198] J. J. McGregor, P. Willett, *J. Chem. Inf. Comput. Sci.* **21** (1981) 137–140.

[10.199] P. E. Blower, R. C. Dana in P. Willett (ed.): *Modern Approaches to Chemical Reaction Searching*, Gower, Aldershot 1986, p. 146.

[10.200] P. E. Blower et al. in W. A. Warr (ed.): *Chemical Structures, The International Language of Chemistry*, Springer Verlag, Berlin 1988, p. 399.

[10.201] R. E. Buntrock, *Database* **11** (1988) no. 6, 124–127.

[10.202] A. P. Johnson in W. A. Warr (ed.): *Chemical Structures, The International Language of Chemistry*, Springer Verlag, Berlin 1988, p. 297.

[10.203] A. P. Johnson, A. P. Cook in P. Willett (ed.): *Modern Approaches to Chemical Reaction Searching*, Gower, Aldershot 1986, p. 184.

[10.204] E. Zass, S. Muller, *Chimia* **40** (1986) no. 2, 38–50.

[10.205] J. H. Borkent, F. Oukes, J. H. Noordik, *J. Chem. Inf. Comput. Sci.* **28** (1988) 148–150.

[10.206] A. P. Johnson et al. in W. A. Warr (ed.): *Chemical Structure Information Systems: Interfaces, Communication and Standards*, ACS Symposium Series 400, American Chemical Society, Washington, 1989, p. 50.

[10.207] T. E. Moock, J. G. Nourse, D. Grier, W. D. Hounshell in W. A. Warr (ed.): *Chemical Structures, The International Language of Chemistry*, Springer Verlag, Berlin 1988, p. 303.

[10.208] G. Grethe, D. del Rey, J. G. Jacobson, M. Van Duyne in W. A. Warr (ed.): *Chemical Structures, The International Language of Chemistry*, Springer Verlag, Berlin 1988, p. 315.

[10.209] W. T. Wipke et al. in P. Willett (ed.): *Modern Approaches to Chemical Reaction Searching*, Gower, Aldershot 1988, p. 92.

[10.210] D. F. Chodosh in P. Willett (ed.): *Modern Approaches to Chemical Reaction Searching*, Gower, Aldershot 1986, p. 118.

[10.211] D. Bawden, S. Wood in P. Willett (ed.): *Modern Approaches to Chemical Reaction Searching*, Gower, Aldershot 1986, p. 78.

[10.212] S. Hanessian, F. Major, S. Leger, *New Methods Drug. Res.* **1** (1985) 201–224.

[10.213] H. Gelernter et al., *Science (Washington, D. C.)* **197** (1977) 1041–1049.

[10.214] K. K. Agarwal, D. L. Larsen, H. Gelernter, *Comput. Chem.* **2** (1978) 75–84.

[10.215] J. Gasteiger et al., *Top. Curr. Chem.* **137** (1987) 19–73.

[10.216] J. Dugundji, I. Ugi, *Top. Curr. Chem.* **39** (1973) 19–64.

[10.217] J. Bauer, R. Herges, E. Fontain, I. Ugi, *Chimia* **39** (1985) 43–53.

[10.218] R. Herges in W. A. Warr (ed.): *Chemical Structures, The International Language of Chemistry*, Springer Verlag, Berlin 1988, p. 385.

[10.219] J. B. Hendrickson, *Acc. Chem. Res.* **19** (1986) 274–281.

[10.220] J. B. Hendrickson, A. G. Toczko, *J. Chem. Inf. Comput. Sci.* **29** (1989) 137–145.

[10.221] K. Funatsu, S. I. Sasaki, *Tetrahedron Comput. Methodol.* **1** (1988) 27–37.

References for Chapter 11

[11.1] A. Barr, E. A. Feigenbaum (eds.): *The Handbook of Artificial Intelligence*, vol. I, William Kaufmann, Los Altos, CA, 1981.

[11.2] A. Barr, E. A. Feigenbaum (eds.): *The Handbook of Artificial Intelligence*, vol. II, William Kaufmann, Los Altos, CA, 1982.

[11.3] P. R. Cohen, E. A. Feigenbaum (eds.): *The Handbook of Artificial Intelligence*, vol. III, William Kaufmann, Los Altos, CA, 1982.

[11.4] E. Rich: *Artificial Intelligence*, McGraw-Hill, New York 1983.

[11.5] P. H. Winston: *Artificial Intelligence*, 2nd ed., Addison-Wesley, Reading 1984.

[11.6] L. F. Lunin, L. C. Smith, *J. Am. Soc. Inf. Sci.* **35** (1984) no. 5, 278–279.

[11.7] N. Cercone, G. McCalla, *J. Am. Soc. Inf. Sci.* **35** (1984) no. 5, 280–290.

[11.8] A. Bonnet: *Artificial Intelligence: Promise and Performance*, Prentice Hall, Hemel Hempstead 1985.

[11.9] S. C. Shapiro (ed.): *Encyclopedia of Artificial Intelligence*, Wiley-Interscience, Chichester 1987.

[11.10] A. M. Turing, *Mind* **59** (1950) no. 236, 433–460.

[11.11] E. A. Feigenbaum, P. McCorduck: *The Fifth Generation: Artificial Intelligence and Japan's Computer Challenge to the World*, Addison-Wesley, Reading 1983.

[11.12] N. K. Herther, *Microcomputers for Information Management* **3** (1986) no. 1, 31–45.

[11.13] P. Davies, *Advanced Information Report*, March 1990, 1–4.

[11.14] P. Langley, J. G. Carbonell, *J. Am. Soc. Inf. Sci.* **35** (1984) no. 5, 306–316.

[11.15] P. Davies: *Artificial Intelligence: Potential for Application in the Information Industry*, EUSIDIC Research Report (EUSIDIC, Calne, Wiltshire, UK) 1988, Knowledge Software Management, 63 College Piece, Mortimer, Reading, UK. An updated version is in the press, Learned Information, Medford NJ.

[11.16] R. C. Johnson, C. Brown: *Cognizers: Neural Networks and Machines That Think*, Wiley, Chichester 1988.

[11.17] A. Colmerauer, H. Kanoui, R. Pasero, Ph. Roussel: "Un Système de Communication Homme-machine en Français," Res. Rep. Groupe Intelligence Artificielle, Faculté des Sciences de Luminy, Marseilles 1973.

[11.18] J. D. Foley, *Sci. Am.* **257** (1987) no. 4, 82–90.

[11.19] R. Bisiani, *Ann. N. Y. Acad. Sci.* **405** (1983) 39–47.

[11.20] F. J. Smith, R. J. Linggard, *Lect. Notes Comp. Sci.* **146** (1983) 275–288.

[11.21] L. D. Erman, F. Hayes-Roth, V. R. Lesser, D. R. Reddy, *Comput. Surveys* **12** (1980) 213–253.

[11.22] D. T. Hawkins; *Online (Weston, Conn.)* **11** (1987) no. 5, 91–98.

[11.23] E. H. Shortliffe: *Computer-Based Medical Consultations: MYCIN*, American Elsevier, New York 1976.
B. G. Buchanan, E. H. Shortliffe (eds.): *Rule-Based Expert Systems: The MYCIN Experiments of the Stanford Heuristic Programming Project*, Addison-Wesley, Reading 1984.

[11.24] D. Gross, *Online Rev.* **12** (1988) no. 5, 283–289.

[11.25] B. Vickery, A. Vickery, *J. Inf. Sci.* **16** (1990) 65–70.

[11.26] A. Vickery, *Aslib Information* **17** (1989) nos. 11/12, 271–274.

[11.27] R. Grishman, *J. Am. Soc. Inf. Sci.* **35** (1984) no. 5, 291–296.

[11.28] *La Traduction Assistée par Ordinateur*, Observatoire des Industries de la Langue, Paris 1989.

[11.29] Proceedings of the Congress on Machine Translation held in Munich, 16–18 August 1989, Deutsche Gesellschaft für Dokumentation, Frankfurt.

[11.30] P. Mayorcas (ed.): *Translating and the Computer 10*, Aslib, London 1989.

[11.31] M. Nagao: *Machine Translation: How Far Can it Go?*, Oxford University Press, Oxford 1989.

[11.32] R. E. Dessy, *Anal. Chem.* **56** (1984) 1200A-1212A.

[11.33] J. K. Kastner, S. J. Hong, *Eur. J. Operational Res.* **18** (1984) no. 3, 285–292.

[11.34] N. S. Yaghmai, J. A. Maxin, *J. Am. Soc. Inf. Sci.* **35** (1984) no. 15, 297–305.

[11.35] R. Forsyth: *Expert Systems: Principles and Case Studies*, Chapman and Hall, London 1984.

[11.36] S. M. Weiss, C. A. Kulikowski: *A Practical Guide to Designing Expert Systems*, Rowman and Allanheld, Lanham, MD, 1984.

[11.37] P. C. Jurs: *Computer Software Applications in Chemistry*, John Wiley and Sons, New York 1986, Chap. 15, p. 212.

[11.38] A. Hart: *Expert Systems: An Introduction for Managers*, Kogan Page, London 1988.

[11.39] M. O'Neill, *Aslib Proc.* **41** (1989) no. 4, 163–168.

[11.40] R. Alberico, M. Micco: *Expert Systems for Reference and Information Retrieval*, Meckler, London 1990.

[11.41] B. A. Hohne, T. H. Pierce (eds.): *Expert System Applications in Chemistry*, ACS Symposium Series 408, American Chemical Society, Washington, DC, 1989.

[11.42] B. M. Carrington, *Database* **13** (1990) no. 2, 47–50.

[11.43] W. F. Clocksin, C. S. Mellish: *Programming in PROLOG*, Springer-Verlag, Berlin 1981.

[11.44] C. Paice, *Aslib Proc.* **38** (1986) no. 10, 343–353.

[11.45] N. N. Mitev, S. Walker in: *Advances in Intelligent Retrieval: Informatics 8*, Aslib, London 1986, p. 215.

[11.46] A. Kemp: *Computer-Based Knowledge Retrieval*, Aslib, London 1988.

[11.47] N. Harrison, B. Murphy, *Library HiTech.* **19** (1987) no. 5, 77–80.

[11.48] D. T. Hawkins, *Online (Weston, Conn.)* **12** (1988) no. 1, 31–43.

[11.49] A. Vickery, H. M Brooks, B. Robinson, *J. Doc.* **43** (1987) no. 1, 1–23.

[11.50] A. Vickery, H. M. Brooks, *Information Processing and Management* **23** (1987) 99–117.

[11.51] J. Hares, M. Thomas, *Computing*, June 1988, 14–17.

[11.52] S. Wade et al., *Online Rev.* **12** (1988) no. 2, 91–108.

[11.53] I. G. Hendry, P. Willett, F. E. Wood, *Program* **20** (1986) 245–263.

[11.54] I. G. Hendry, P. Willett, F. E. Wood, *Program* **20** (1986) 382–393.

[11.55] S. J. Wade, P. Willett, *Program* **22** (1988) 44–61.

[11.56] S. J. Wade, P. Willett, D. Bawden, *J. Inf. Sci.* **15** (1989) 249–260.

[11.57] N. J. Belkin, H. M. Brooks, P. J. Daniels, *Int. J. of Man-Machine Studies* **27** (1987) no. 2, 127–144.

[11.58] N. J. Belkin, W. B. Croft, *Annu. Rev. Inf. Sci. Technol.* **22** (1987) 109–146.

[11.59] H. M. Brooks, N. J. Belkin, P. J. Daniels in: *Advances in Intelligent Retrieval: Informatics 8*, Aslib, London 1986, p. 191.

[11.60] H. M. Brooks, P. J. Daniels, N. J. Belkin, *Journal of Information and Image Management* **12** (1986) nos. 1/2, 37–44.

[11.61] I. Wormell (ed.): *Knowledge Engineering: Expert Systems and Information Retrieval*, Taylor Graham, London 1987.

[11.62] M. O'Neill, A. Morris, *The Electronic Library* **7** (1989) no. 5, 295–300.

[11.63] E. J. Corey, A. K. Long, S. D. Rubenstein, *Science (Washington, D. C.)* **228** (1985) 408–418.

[11.64] D. A. Pensak, E. J. Corey in W. T. Wipke, W. J. Howe (eds.): *Computer-Assisted Organic Synthesis*, ACS Symposium Series 61, American Chemical Society, Washington, DC, 1977, p. 1.

[11.65] W. T. Wipke et al. in W. T. Wipke, W. J. Howe (eds.): *Computer-Assisted Organic Synthesis*, ACS Symposium Series 61, American Chemical Society Washington, DC, 1977, p. 97.

[11.66] W. T. Wipke, G. I. Ouchi, S. Krishnan, *Artificial Intelligence* **11** (1978) 173–193.

[11.67] P. Gund et al., *J. Chem. Inf. Comput. Sci.* **20** (1980) 88–93.

[11.68] M. Bersohn, A. Esack, *Chem. Rev.* **76** (1976) 269–282.

[11.69] J. Gasteiger, M. G. Hutchings, P. Low, H. Saller in T. H. Pierce, B. A. Hohne (eds.): *Artificial Intelligence Applications in Chemistry*, ACS Symposium Series 306, American Chemical Society, Washington, DC, 1986, p. 258.

[11.70] J. Gasteiger, P. Röse, H. Saller, *J. Mol. Graphics* **6** (1988) no. 2, 87–97.

[11.71] W. T. Wipke, W. J. Howe (eds.): *Computer-Assisted Organic Synthesis*, ACS Symposium Series 61, American Chemical Society, Washington, DC, 1977.

[11.72] W. T. Wipke, D. P. Dolata in T. H. Pierce, B. A. Hohne (eds.): *Artificial Intelligence Applications in Chemistry*, ACS Symposium Series 306, American Chemical Society, Washington, DC, 1986, p. 188.

[11.73] A. J. Gushurst, W. L. Jorgensen, *J. Org. Chem.* **53** (1988) no. 15, 3397–3408.

[11.74] H. Gelernter et al., *Science (Washington, D. C.)* **197** (1977) 1041–1049.

[11.75] K. Funatsu, S. Sasaki, *Tetrahedron Comput. Methodol.* **1** (1988) no. 1, 39–51.

[11.76] J. B. Hendrickson et al. in B. A. Hohne, T. H. Pierce (eds.): *Expert System Applications in Chemistry*, ACS Symposium Series 408, American Chemical Society, Washington 1989, p. 62.

[11.77] D. H. Smith (ed.): *Computer-Assisted Structure Elucidation*, ACS Symposium Series 54, American Chemical Society, Washington, DC, 1977.

[11.78] W. Bremser et al.: *Carbon-13 NMR Spectral Data*, 3rd ed., Verlag Chemie, Weinheim 1981.

[11.79] R. E. Dessy, *Anal. Chem.* **56** (1984) 1312A-1332A.

[11.80] N. A. B. Gray: *Computer-Assisted Structure Elucidation*, John Wiley, New York 1986.

[11.81] N. A. B. Gray, *Anal. Chim. Acta* **210** (1988) 9–32.

[11.82] J. Zupan (ed.): *Computer-Supported Spectroscopic Databases*, Ellis–Horwood, Chichester 1986.

[11.83] T. P. Bridge, M. H. Williams, A. F. Fell, *Chem. Br.*, Nov. 1987, 1085–1088.

[11.84] P. C. Jurs: *Computer Software Applications in Chemistry*, John Wiley, New York 1986, Chap. 16, p. 219.

[11.85] B. G. Buchanan, E. A. Feigenbaum, *Artificial Intelligence* **11** (1978) 5–24.

[11.86] R. K. Lindsay, B. G. Buchanan, E. A. Feigenbaum, J. Lederberg: *Applications of Artificial Intelligence for Organic Chemistry; the DENDRAL Project*, McGraw-Hill, New York 1980.

[11.87] N. A. B. Gray, *Prog. Nucl. Magn. Res. Spectrosc.* **15** (1982) 201–248.

[11.88] M. Passlack, W. Bremser in J. Zupan (ed.): *Computer-Supported Spectroscopic Databases*, Ellis–Horwood, Chichester 1986, p. 92.

[11.89] W. Bremser, R. Neudert, *Eur. Spectrosc. News* **75** (1987) 10–27.

[11.90] R. Neudert, W. Bremser, H. Wagner, *Org. Mass Spectrom.* **22** (1987) 321–329.

[11.91] W. Bremser, *Angew. Chem. Int. Ed. Engl.* **27** (1988) 247–260.

[11.92] S. Sasaki, Y. Kudo, *J. Chem. Inf. Comput. Sci.* **25** (1985) 252–257.

[11.93] K. Funatsu, N. Miyabayashi, S. Sasaki, *J. Chem. Inf. Comput. Sci.* **28** (1988) 18–28.

[11.94] K. Funatsu, Y. Susuta, S. Sasaki, *J. Chem. Inf. Comput. Sci.* **29** (1989) 6–17.

[11.95] H. B. Woodruff, *Trends Anal. Chem.* **3** (1984) 72–75.

[11.96] H. B. Woodruff, S. A. Tomellini, G. M. Smith in T. H. Pierce, B. A. Hohne (eds.): *Artificial Intelligence Applications in Chemistry*, ACS Symposium Series 306, American Chemical Society, Washington, DC, 1986, p. 312.

[11.97] L. S. Ying, S. P. Levine, S. A. Tomellini, S. R. Lowry, *Anal. Chim. Acta* **210** (1988) 51–62.

[11.98] F. W. McLafferty, D. B. Stauffer, *J. Chem. Inf. Comput. Sci.* **25** (1985) 245–252.

[11.99] C. A. Shelley, M. E. Munk, *Anal. Chim. Acta* **133** (1981) 507–516.

[11.100] M. E. Munk, M. Farkas, A. Lipkis, B. D. Christie, *Mikrochim. Acta* **1986** II (1987) 199–215.

[11.101] L. A. Gribov, *Anal. Chim. Acta* **122** (1980) 249–256.

[11.102] J. Zupan, M. Novic, S. Bohanec, M. Razinger, *Anal. Chim. Acta* **200** (1987) 333–345.

[11.103] H. J. Luinge, J. H. Van der Maas, *Anal. Chim. Acta* **223** (1989) 135–147.

[11.104] H. Kalchhauser, W. Robien, *J. Chem. Inf. Comput. Sci.* **25** (1985) 103–108.

[11.105] C. J. Rawlings, *Biochem. Soc. Trans.* **17** (1989) 851–855.

[11.106] C. Trindle, *J. Mol. Graphics* **6** (1988) no. 2, 67–73.

[11.107] C. Trindle in B. A. Hohne, T. H. Pierce (eds.): *Expert System Applications in Chemistry*, ACS Symposium Series 408, American Chemical Society, Washington, DC, 1989, p. 92.

[11.108] D. P. Dolata, A. R. Leach, K. Prout, *Journal of Computer-Aided Molecular Design* **1** (1987) 73–85.

[11.109] D. P. Dolata, R. E. Carter, *J. Chem. Inf. Comput. Sci.* **27** (1987) 36–47.

[11.110] D. P. Dolata, A. R. Leach, K. Prout in W. G. Richards (ed.): *Computer-Aided Molecular Design*, IBC Technical Services, London 1989, p. 67.

[11.111] W. T. Wipke, M. A. Hahn, *Tetrahedron Comput. Methodol.* **1** (1988) 141–167.

[11.112] M. A. Hahn, W. T. Wipke in W. A. Warr (ed.): *Chemical Structures, The International Language of Chemistry*, Springer Verlag, Berlin 1988, p. 269.

[11.113] F. Darvas, *J. Mol. Graphics* **6** (1988) no. 2, 80–86.

[11.114] T. Koschmann et al., *J. Mol. Graphics* **6** (1988) no. 2, 74–79.

[11.115] T. H. Pierce, B. A. Hohne (eds.): *Artificial Intelligence Applications in Chemistry*, ACS Symposium Series 306, American Chemical Society, Washington, DC, 1986.

[11.116] J. M. Hushon (ed.): *Expert Systems for Environmental Applications*, ACS Symposium Series 431, American Chemical Society, Washington, DC, 1990.

[11.117] C.-C. Chen, *Microcomputers for Information Management* **6** (1989) no. 2, 77–97.

[11.118] C.-C. Chen, *Microcomputers for Information Management* **6** (1989) no. 2, 135–145.

[11.119] L. Davenport, B. Cronin, *J. Inf. Sci.* **15** (1989) 369–372.

[11.120] C. Franklin, *Online (Weston, Conn.)* **13** (1989) no. 3, 37–49.

[11.121] *Commun. ACM* **31** (1988), special issue on hypertext.

[11.122] *J. Am. Soc. Inf. Sci.* **40** (1989) no. 3, special volume on hypertext.

[11.123] V. Bush, *The Atlantic Monthly*, July 1945, 101–108.

[11.124] V. Bush in: *Science is Not Enough*, Apollo Editions, New York 1969, p. 75.

[11.125] T. H. Nelson: *Literary Machines*, Tempus Press (Microsoft), Tell City, IN, 1987.

References for Chapter 12

[12.1] W. Dijkhuis, *J. Inf. Sci.* **4** (1982) 175.
[12.2] M. W. Hill in: B. T. Stern (ed.), *Information and Innovation.* North Holland Publ., Amsterdam 1982.
[12.3] A. J. Harrison, *Science* **223** (1984) 543.
[12.4] F. R. Bradbury, M. C. McCarthy, C. W. Suckling, *Chem. Ind. London* **1972** 22, 105, 195.
[12.5] S. Greif, *Naturwissenschaften* **76** (1989) 156.
[12.6] J. Phillips, *Endeavour* **8** (1984) 90.
[12.7] F. Machlup, *GRUR Int.* **63** (1961) 373, 473.
[12.8] C. Oppenheim, *J. Chem. Inf. Comput. Sci.* **18** (1978) 56.
[12.9] J. W. Baxter: *World Patent Law and Practice*, Sweet & Maxwell, London, Matthew Bender & Co, New York 1968.
[12.10] R. Calvert (ed.): *Encyclopedia of Patent Practice and Innovation Management*, Robert E. Krieger Publ., Huntingdon, NY, 1974.
[12.11] H. Schade, J. Schade: *Patents at a Glance*, 3rd ed., Carl Heymanns Verlag, Cologne-Berlin-Bonn-Munich, 1980.
[12.12] European Federation of Pharmaceutical Industries' Associations: *Memorandum on the need of the European Pharmaceutical Industry for restoration of effective patent term for pharmaceuticals* (without issue date).
[12.13] J. Morehead, *Tech. Serv. Quart. (New York)* **3** (1986) 339.
[12.14] World Intellectual Property Organization: *WIPO Publication no.* 400, Geneva 1990.
[12.15] *Ind. Property*, March 1991, p. 182.
[12.16] European Patent Office: *How to get a European Patent*, Munich (updated editions appear as appropriate).
[12.17] European Patent Office: *Annual Report 1990*, chapter IV., Munich 1991.
[12.18] S. Wyatt, G. Bertin, K. Pavitt, *World Pat. Inf.* **7** (1985) 196.
[12.19] F. Lederer, K. H. Oppenländer, G. Eisenführ, I. Traub, V. Deneke, *Mitt. Dtsch. Patentanw.* **76** (1985) 101.
[12.20] K. H. Oppenländer, *Ind. Property* **25** (1986) 494.
[12.21] E. Lins, M. Rau, *Mitt. Dtsch. Patentanw.* **81** (1990) 114.
[12.22] J. Stephenson, *World Pat. Inf.* **4** (1982) 164.

References for Chapter 13

[13.1] W. H. Bowman, *J. Chem. Inf. Comput. Sci.* **18** (1978) 81.
[13.2] S. M. Kaback, *CHEMTECH* **10** (1980) 172.
[13.3] F. Liebesny, J. W. Hewitt, P. S. Hunter, M. Hannah, *Inf. Sci. London* **1974**, 169.
[13.4] H. Mann, A. Hellyer, *World Pat. Inf.* **2** (1980) 27.
[13.5] J. Allen, C. Oppenheim, *World Pat. Inf.* **2** (1980) 77.
[13.6] E. S. Simmons, *J. Chem. Inf. Comput. Sci.* **25** (1985) 379.
[13.7] P. Newman, E. I. Hoegberg, *J. Chem. Inf. Comput. Sci.* **18** (1978) 8.

References for Chapter 14

[14.1] Chemical Abstracts Service: *Statistical Summary 1907–1988*, Columbus, Ohio, May 1989, CAS 1433.

References for Chapter 15

[15.1] The Crown Copyright (never exercised) on British patent documents was abolished when the Patents Act 1977 came into force.

References for Chapter 16

[16.1] World Intellectual Property Organization: *Handbook on Industrial Property Information and Documentation*, vol. I, Part 3.

[16.2] The expression „Markush formulae" refers to the decision ex parte Markush, 1925 C. D. 126, 340 OG 839, allowing, for the first time, the wording: „a member selected from the group consisting of ..." instead of „or", which was not deemed acceptable when used in US patent claims. To apply this expression to generic structural formulae is mistaken, for at least three reasons: (1) Many Markush type claims cannot be represented as generic structural formulae (neither can the claims of Eugene A. Markush's US patent 1 506 316); there is no one-to-one relationship of Markush claims and generic structural formulae. (2) The decision ex parte Markush pertains only to the practice of the United States Patent and Trademark Office, it has no international significance. (3) Generic structural formulae abound in (non-US) patent documents published long before 1925, e.g., in German patent documents issued in the 1880s; see reference [19.59].

References for Chapter 17

[17.1] S. Hahnemann, *World Pat. Inf.* **4** (1982) 172.

[17.2] *Off. J. EPO* **3** (1990) 81.

References for Chapter 18

[18.1] Kenneth Mason Publ. Ltd, Elmsworth, England.

[18.2] International Technical Disclosure Inc., Tinley Park, IL, USA.

[18.3] VCH Verlagsgesellschaft, Weinheim, Germany.

[18.4] Carl Heymanns Verlag, Cologne-Berlin-Bonn-Munich.

[18.5] Patent and Trademark Office Society, Arlington, VA, USA.

[18.6] Pergamon Press, Oxford-New York-Frankfurt-Tokyo-Sidney.

[18.7] WILA-Verlag für Wirtschaftswerbung Wilhelm Lampl, Munich.

References for Chapter 19

[19.1] A telling example: In a series of remarkable decisions taken by the District Court, District of Massachusetts (No. 76-1634 Z, decided September 13, 1985, and October 11, 1985; 228 USPQ 305), and the Court of Appeals, Federal Circuit (No. 86-604, decided January 7 and April 25, 1986; 229 USPQ 561, 5 USPQ 2d 1080), Eastman Kodak Company was held guilty of infringement of several Polaroid Corporation patents and sentenced to pay a very high sum of money as well as closing down a branch of its business.

[19.2] S. R. Heller, G. W. A. Milne, R. J. Feldmann, *J. Chem. Inf. Comput. Sci.* **16** (1976) 232.

[19.3] B. Charton, *J. Chem. Inf. Comput. Sci.* **17** (1977) 45.

[19.4] A. Nevyjel, O. Oberhauser, Publication. no. 4241, Österreichisches Forschungszentrum Seibersdorf, Oct. 1983.

[19.5] S. M. Kaback, *J. Chem. Inf. Comput. Sci.* **24** (1984) 159.

[19.6] S. M. Kaback, *J. Chem. Inf. Comput. Sci.* **25** (1985) 371.

[19.7] H. D. White, B. C. Griffith, *Inf. Process. Manage.* **23** (1987) 211.

[19.8] U. Schoch-Grübler, *Online Rev.* **14** (1990) 95.

[19.9] C. V. Cleverdon, *J. Doc.* **28** (1972) 195.

[19.10] D. R. Swanson, *Libr. Q.* **41** (1971) 223.

[19.11] R. Fugmann, *Int. Classif.* **9** (1982) 140.

[19.12] H. H. Wellisch, *Database* **6** (1983) 54.

[19.13] World Intellectual Property Organization: *Standardisation Among Computerized Search Systems Having Substantial Coverage of Patent Documents*, Document prepared by the International Bureau, PIF/86/RT/1, June 7, 1986.

[19.14] T. M. Johns, W. G. Andrus, S. de Voe, J. M. Myers, R. G. Smith, O. Uhlir, *J. Chem. Inf. Comput. Sci.* **19** (1979) 241.

[19.15] E. S. Simmons, *Database* **8** (1985) 49.

[19.16] An instructive but not exceptional example is the US patent specification no. 4 215 144 (July 29, 1980) citing on its title page 58 application numbers.

[19.17] E. Garfield, *Naturwissenschaften* **68** (1981) 519.

[19.18] M. A. Lobeck, *Nachr. Dok.* **32** (1981) 20.

[19.19] L. van Bommel, *World Pat. Inf.* **5** (1983) 35.

[19.20] K. Faust: „*Früherkennung technischer Entwicklungen auf der Basis von Patentdaten,*" *IfO – Studien zur Strukturforschung* **9**, 2 vol. IfO – Institut für Wirtschaftsförderung, Munich 1987.

[19.21] H. Aspden, *World Pat. Inf.* **5** (1983) 170.

[19.22] J. E. Hibbs, R. F. Bobner, I. Newman, C. M. Dye, C. R. Benz, *Online (Weston, CT)* **8** (1984) 59.

[19.23] D.-M. Harmsen, *Innovation* **1986**, no. 2, 162.

[19.24] M. Moureau, A. Girard in: M. E. Williams, T. H. Hogan (eds.): *Proceedings of the National Online Meeting*, New York, May 5–7, 1987, p. 351. Learned Information Inc., Medford, NJ.

[19.25] K. W. M. Tyson, *Online (Weston, CT)* **12** (1988) 85.

[19.26] World Intellectual Property Organization: *Handbook on Industrial Property Information and Documentation*, Vol. I, Standard ST14.

[19.27] S. E. Robertson, *J. Doc.* **33** (1977) 294.

[19.28] S. E. Robertson, M. E. Maron, W. S. Cooper, *Inf. Tech. Res. Dev.* **1** (1982) 1.

[19.29] S. E. Robertson, *J. Doc.* **42** (1986) 182.

[19.30] L. Schamber, M. B. Eisenberg, M. S. Nilan, *Inf. Process. Manage.* **26** (1990) 755.

[19.31] T. Saracevic, *J. Am. Soc. Inf. Sci.* **26** (1975) 321.

[19.32] A. M. Rees, *Aslib Proc.* **18** (1966) 316.

[19.33] C. W. N. Thompson, *J. Am. Soc. Inf. Sci.* **24** (1973) 270.

[19.34] R. J. Rowlett, Jr., *J. Chem. Inf. Comput. Sci.* **25** (1985) 159.

[19.35] The Institute of Information Scientists, *Searcher – The Newsletter of the Patent & Trademark Group* no. 56, June 1991, p. 5.

[19.36] P. Claus, *World Pat. Inf.* **2** (1980) 13.

[19.37] World Intellectual Property Organization: *Handbook on Industrial Property Information and Documentation*, vol. II, part 5.

[19.38] Zheng Fen, *World Pat. Inf.* **9** (1987) 152.

[19.39] S. de Vries, *World Pat. Inf.* **11** (1989) 115 .

[19.40] G. Jenkins, *World Pat. Inf.* **11** (1989) 121, 187.

[19.41] K. H. Ullmer, *9th Meeting of the CIDST Working Group*: 'Patents Documentation' (June 7, 1976). Commission of the European Communities, Directorate General Scientific and Technical Information and Information Management. Doc. 3380/76, Luxembourg, Nov. 22, 1976.

[19.42] A. M. Carpenter, M. Jones, C. Oppenheim, *Int. Classif.* **5** (1978) 30.

[19.43] T. S. Eisenschitz, C. Oppenheim, *Int. Classif.* **6** (1979) 26.

[19.44] M. F. Lynch, J. M. Barnard, S. M. Welford, *J. Chem. Inf. Comput. Sci.* **21** (1981) 148.

[19.45] M. F. Lynch in J. M. Barnard (ed.): *Computer Handling of Generic Chemical Structures*, Gower Pub. Co., Aldershot-Brookfield 1984, p. 1.

[19.46] E. Meyer, P. Schilling, E. Sens in J. M. Barnard (ed.): *Computer Handling of Generic Chemical Structures*, Gower Publ. Co., Aldershot-Brookfield 1984, p. 83.

[19.47] W. Fisanick in J. M. Barnard (ed.): *Computer Handling of Generic Chemical Structures*, Gower Pub. Co., Aldershot-Brookfield 1984, p. 106.

[19.48] S. M. Welford, S. Ash, J. M. Barnard, L. Carruthers, M. F. Lynch, A. von Scholley in J. M. Barnard (ed.): *Computer Handling of Generic Chemical Structures*, Gower Pub. Co., Aldershot-Brookfield 1984, p. 130.

[19.49] G. M. Downs, V. J. Gillet, J. Holliday, M. F. Lynch in W. A. Warr (ed.): *Chemical Structures. The International Language of Chemistry*, Springer Verlag, Berlin-Heidelberg-New York-London-Paris-Tokyo 1988, p. 151.

[19.50] K. E. Shenton, P. Norton, E. A. Ferns in: W. A. Warr (ed.), *Chemical Structures. The International Language of Chemistry*, Springer Verlag, Berlin-Heidelberg-New York-London-Paris-Tokyo 1988, p. 169.

[19.51] R. Fugmann in W. A. Warr (ed.): *Chemical Structures. The International Language of Chemistry*, Springer Verlag, Berlin-Heidelberg-New York-London-Paris-Tokyo 1988, p. 425.

[19.52] E. Meyer, *J. Chem. Inf. Comput. Sci.* **31** (1991) 68.

[19.53] C. Suhr, W. Dethlefsen, *9th International Online Information Meeting*, London December 3–5, 1985.

[19.54] W. Dethlefsen, V. J. Gillet, G. M. Downs, J. D. Holliday, J. M. Barnard, *J. Chem. Inf. Comput. Sci.* **31** (1991) 233.

[19.55] W. Dethlefsen, V. J. Gillet, G. M. Downs, J. D. Holliday, J. Barnard, *J. Chem. Inf. Comput. Sci.* **31** (1991) 253.

[19.56] J. M. Barnard, *Database* **10** (1987) 27.

[19.57] E. S. Simmons, *J. Chem. Inf. Comput. Sci.* **31** (1991) 45.

[19.58] K. E. H. Göhring, J. F. Sibley, *World Pat. Inf.* **11** (1989) 5.

[19.59] C. Suhr, *Proceedings of the International Chemical Information Conference*, Montreux, September 26–28, 1989, p. 131, Infonortics Ltd., Calne 1989.

[19.60] F. A. Jenny, *World Pat. Inf.* **12** (1990) 71.

[19.61] J. F. Sibley, *J. Chem. Inf. Comput. Sci.* **31** (1991) 5.

[19.62] G. W. A. Milne, *J. Chem. Inf. Comput. Sci.* **31** (1991) 9.

[19.63] F. W. Lancaster, *Adv. Librarianship* **7** (1977) 1.

[19.64] Chai Kim, *J. Am. Soc. Inf. Sci.* **24** (1973) 148.

[19.65] E. Meyer, R. Jansen, E. Sens, *Nachr. Dok.* **23** (1972) 203.

[19.66] R. Fugmann, *Int. Classif.* **1** (1974) 76.

[19.67] C. Martinez, J. Lucey, E. Linder, *J. Chem. Inf. Comput. Sci.* **27** (1987) 158.

[19.68] C. Hansen Fenichel, *Libr. Res.* **2** (1980/1981) 107.

[19.69] D. C. Blair, M. E. Maron, *Commun. ACM* **28** (1985) 289.

[19.70] D. C. Blair, *Int. Classif.* **13** (1986) 18.

[19.71] D. C. Blair, M. E. Maron, *Inf. Process. Manage.* **26** (1990) 437.

[19.72] R. Wagers, *Online (Weston, Conn.)* **7** (1983) 60.

[19.73] W. J. Maykrantz, *Inf. Serv. Use* **8** (1988) 257.

[19.74] A. Wittmann, L. Schikarski, *Mitt. Dtsch. Patentanw.* **75** (1984) 221.

References for Chapter 20

[20.1] World Intellectual Property Organization: *Handbook on Industrial Property Information and Documentation*, vol. II, part 6.

[20.2] J. Foulkes (ed.): *Downloading Bibliographic Records*, Proceedings of a One-Day Seminar Sponsored by the MARC Users' Group, Gower Publ. Co., Aldershot 1986.

[20.3] A. E. Jackson, *Aslib Proc.* **40** (1988) 111.

[20.4] P. J. Pollick, *J. Chem. Inf. Comput. Sci.* **23** (1983) 160.

[20.5] W. Fisanick, *J. Chem. Inf. Comput. Sci.* **30** (1990) 145.

[20.6] T. Ebe, K. A. Sanderson, P. S. Wilson, *J. Chem. Inf. Comput. Sci.* **31** (1991) 31 .

[20.7] S. M. Kaback, *J. Chem. Inf. Comput. Sci.* **17** (1977) 143.

[20.8] S. M. Kaback, *J. Chem. Inf. Comput. Sci.* **20** (1980) 1.

[20.9] M. D. Dixon, C. Oppenheim, *World Pat. Inf.* **4** (1982) 60.

[20.10] C. Oppenheim, *Sci. Technol. Libr.* **2** (1982).

[20.11] R. Fugmann, *J. Chem. Inf. Comput. Sci.* **22** (1982) 117, 118.

[20.12] R. Fugmann, *Nachr. Dok.* **14** (1963) 179.

[20.13] S. Rössler, A. Kolb, *J. Chem. Doc.* **10** (1970) 138.

[20.14] M. A. Lobeck, *Nachr. Dok.* **25** (1974) 210.

[20.15] F. Ehrhardt, H. Roschkowski, *J. Chem. Inf. Comput. Sci.* **26** (1986) 63.

[20.16] R. Fugmann, H. Nickelsen, I. Nickelsen, J. H. Winter, *J. Am. Soc. Inf. Sci.* **25** (1974) 282.

[20.17] C. Suhr, *Chem. Ztg.* **96** (1972) 342.

[20.18] R. Fugmann, *J. Chem. Inf. Comput. Sci.* **19** (1979) 64.

[20.19] E. Meyer, *J. Chem. Doc.* **9** (1969) 109.

[20.20] K. H. Franzreb, P. Hornbach, C. Pahde, G. Ploss, J. Sander, *J. Chem. Inf. Comput. Sci.* **31** (1991) 284.

[20.21] A. Bogsch, *World Pat. Inf.* **10** (1988) 171.

[20.22] W. Pilch, W. Wratschko, *J. Chem. Inf. Comput. Sci.* **18** (1978) 69.

[20.23] W. Pilch, *World Pat. Inf.* **2** (1980) 69.

[20.24] W. Pilch, G. Vacek, *World Pat. Inf.* **10** (1988) 20.

[20.25] R. Fugmann, *Inf. Storage. Retr.* **9** (1973) 353.

[20.26] C. Oppenheim, E. A. Sutherland, *J. Chem. Inf. Comput. Sci.* **18** (1978) 122.

[20.27] R. Bois, J. Chaumier, *World Pat. Inf.* **2** (1980) 61.

[20.28] S. M. Kaback, *World Pat. Inf.* **6** (1984) 36.

[20.29] D. Bawden, J. D. Fisher, *J. Chem. Inf. Comput. Sci.* **25** (1985) 36.

[20.30] T. Novak, *World Pat. Inf.* **9** (1987) 222.

[20.31] S. M. Kaback in J. M. Barnard (ed.): *Computer Handling of Generic Chemical Structures*, Gower Publ. Co., Aldershot-Brookfield 1984, p. 49.

[20.32] S. Martin, G. Bergerhoff, *J. Chem. Inf. Comput. Sci.* **31** (1991) 147.

[20.33] K. E. Cloutier, *J. Chem. Inf. Comput. Sci.* **31** (1991) 40.

[20.34] N. R. Schmuff, *J. Chem. Inf. Comput. Sci.* **31** (1991) 53.

[20.35] C. Suhr, E. von Harsdorf, W. Dethlefsen in J. M. Barnard (ed.): *Computer Handling of Generic Chemical Structures*, Gower Publ. Co., Aldershot-Brookfield 1984, p.96.

[20.36] R. N. Wilke, *J. Chem. Inf. Comput. Sci.* **31** (1991) 36.

[20.37] E. Hyde in S. R. Heller, R. Potenzone (eds.): *Computer Applications in Chemistry*, Elsevier, Amsterdam 1983, p. 1.

[20.38] W. A. Warr, A. R. Haygarth Jackson, *J. Chem. Inf. Comput. Sci.* **28** (1988) 68.

References for Chapter 21

[21.1] G. Turpie, *Database* **11** (1988) 63.

[21.2] United States Patent and Trademark Office, *BNA Patent, Trademark Copyright J.* **40** (1990) 489.
[21.3] M. N. Meller, *Am. Pat. Law Assoc. Q. J.* **11** (1983) double issue nos. 1 & 2.
[21.4] L. Liddle, 1085 OG 6 *Official Gazette*, December 1, 1987).
[21.5] C. Oppenheim, *World Pat. Inf.* **8** (1986) 185.
[21.6] A. Wittmann, G. Tittlbach, *World Pat. Inf.* **8** (1986) 29.
[21.7] European Patent Office: *Search and Documentation in the EPO.* 4th edition, Den Haag 1988.
[21.8] *Off. J. EPO* **12** (1987) 540.

References for Chapter 22

[22.1] C. Suhr, *Chem. Ing. Tech.* **39** (1967) 1299.
[22.2] P. Ochsenbein, *World Pat. Inf.* **9** (1987) 92.
[22.3] Meeting Report, *World Pat. Inf.* **11** (1989) 48.

References for Chapter 23

[23.1] R. L. Wigington, *J. Chem. Inf. Comput. Sci.* **27** (1987) 51.

25 Index